SPACE WEATHER

NATIONAL STRATEGY, ACTION PLAN, AND OBSERVING SYSTEMS FOR PREPAREDNESS

(WITH ACCOMPANYING CD-ROM)

SPACE SCIENCE, EXPLORATION AND POLICIES

Additional books in this series can be found on Nova's website under the Series tab.

Additional e-books in this series can be found on Nova's website under the e-book tab.

SPACE WEATHER

NATIONAL STRATEGY, ACTION PLAN, AND OBSERVING SYSTEMS FOR PREPAREDNESS

(WITH ACCOMPANYING CD-ROM)

PETER BURTON
EDITOR

publishers
New York

For further questions about using the service on copyright.com, please contact:
Copyright Clearance Center
Phone: +1-(978) 750-8400   Fax: +1-(978) 750-4470   E-mail: info@copyright.com.

NOTICE TO THE READER

Library of Congress Cataloging-in-Publication Data

ISBN: 978-1-63484-440-6

Published by Nova Science Publishers, Inc. † New York

CONTENTS

PREFACE

Space weather refers to variations in the space environment between the sun and Earth (and throughout the solar system) that can affect technologies in space and on Earth. Space weather can disrupt the technology that forms the backbone of this country's economic vitality and national security, including satellite and airline operations, communications networks, navigation systems, and the electric power grid. As the Nation becomes ever more dependent on these technologies, space weather poses an increasing risk to infrastructure and the economy. Further, the Strategic National Risk Assessment has identified space weather as a hazard that poses significant risk to the security of the Nation. Clearly, reducing vulnerability to space weather needs to be a national priority. The National Space Weather Strategy (Strategy) and the accompanying National Space Weather Action Plan (Action Plan) together seek to enhance the integration of existing national efforts and to add important capabilities to help meet growing demands for space-weather information. The Strategy and Action Plan build on recent efforts to reduce risks associated with natural hazards and improve resilience of essential facilities and systems, aiming to foster a collaborative environment in which government, industry, and the American people can better understand and prepare for the effects of space weather. The Nation must continue to leverage existing public and private networks of expertise and capabilities and pursue targeted enhancements to improve the ability to manage risks associated with space weather. This book outlines objectives for enhancing the Nation's space-weather readiness in three key areas: national preparedness, forecasting, and understanding. It also describes the study process, the study requirements and their relevance and importance, an assessment and accounting of current and planned space weather observing systems used or to be used for operations, an analysis of gaps between the observing systems' capabilities and their ability to meet documented requirements, and a summary of key findings.

In: Space Weather   ISBN: 978-1-63484-440-6
Editor: Peter Burton   © 2016 Nova Science Publishers, Inc.

Chapter 1

NATIONAL SPACE WEATHER STRATEGY[*]

Space Weather Operations, Research, and Mitigation Task Force

EXECUTIVE OFFICE OF THE PRESIDENT
NATIONAL SCIENCE AND TECHNOLOGY COUNCIL
WASHINGTON, D.C. 20502

October 29, 2015

Dear Colleagues,

Space weather is a naturally occurring phenomenon that has the potential to cause substantial detrimental effects on the Nation's economic and social well-being. Preparing for and predicting space-weather events and their potential effects on Earth is a significant challenge. Recent efforts led by the United States and its international partners have resulted in significant progress toward improving the understanding, monitoring, prediction, and mitigation of this hazard, but much more needs to be done.

Over the past 5 years, OSTP has coordinated interagency efforts to improve the Nation's ability to prepare, avoid, mitigate, respond to, and recover from the potentially devastating impacts of space-weather events.

---

[*] This is an edited, reformatted and augmented version of a document issued by the Office of Science and Technology Policy, National Science and Technology Council, October 2015.

These efforts included the establishment of the interagency Space Weather Operations, Research, and Mitigation (SWORM) Task Force in November 2014. The goal of the SWORM Task Force was to unite the national- and homeland-security enterprise with the science and technology enterprise to formulate a cohesive vision to enhance national preparedness for space weather.

This *National Space Weather Strategy* and accompanying *National Space Weather Action Plan* are the result of the SWORM Task Force's efforts. These documents transcend agency-mission and sector boundaries to describe how the Federal Government will coordinate its efforts on space weather and how the Federal Government plans to engage academia, the private and public sectors, and other governments on space weather. The Strategy and associated Action Plan aim to enhance the preparedness of the Nation by interweaving and building upon existing policy efforts to identify overarching goals that underpin and drive the activities necessary to improve the security and resilience of critical technologies and infrastructures.

These documents represent only a next step to improving national preparedness for space weather. The Strategy sets overall goals for Federal action, while the Action Plan establishes Federal actions and timelines for implementation. Many of these activities will require long time horizons, which will necessitate sustained engagement among government agencies and the private sector. This challenge requires the Nation to work together to continually improve understanding, prediction, and preparedness to enhance the Nation's resilience against severe space-weather events.

Sincerely,
John P. Holdren
Assistant to the President for Science and Technology Director,
Office of Science and Technology Policy

EXECUTIVE SUMMARY

Space weather refers to variations in the space environment between the sun and Earth (and throughout the solar system) that can affect technologies in space and on Earth. Space weather can disrupt the technology that forms the backbone of this country's economic vitality and national security, including satellite and airline operations, communications networks, navigation systems, and the electric power grid. As the Nation becomes ever more dependent on these technologies, space weather poses an increasing risk to infrastructure and the economy. Further, the Strategic National Risk Assessment[1] has identified space weather as a hazard that poses significant risk to the security of the Nation. Clearly, reducing vulnerability to space weather needs to be a national priority.

The *National Space Weather Strategy* (Strategy) and the accompanying *National Space Weather Action Plan* (Action Plan) together seek to enhance the integration of existing national efforts and to add important capabilities to help meet growing demands for space-weather information. The Strategy and Action Plan build on recent efforts to reduce risks associated with natural hazards and improve resilience of essential facilities and systems,[2] aiming to foster a collaborative environment in which government, industry, and the American people can better understand and prepare for the effects of space weather. The Nation must continue to leverage existing public and private networks of expertise and capabilities and pursue targeted enhancements to improve the ability to manage risks associated with space weather.

Six strategic goals underpin the effort to reduce the Nation's vulnerability to space weather:

1) **Establish Benchmarks for Space-Weather Events:** Effective and appropriate actions for space-weather events require an understanding of the magnitude and frequency of such events. Benchmarks will help government and industry assess the vulnerability of critical infrastructure, establish decision points and thresholds for action, understand risk, and provide points of reference to enable mitigation procedures and practices and to enhance response and recovery planning.

2) **Enhance Response and Recovery Capabilities:** There is a need to develop comprehensive guidance to support and improve response and recovery capabilities to manage space-weather events, including the capabilities of Federal, State, and local governments[3] and of the private sector.

3) **Improve Protection and Mitigation Efforts:** Improvements to national preparedness for space-weather events will require enhancing approaches to protection and mitigation. Protection focuses on developing capabilities and actions to secure the Nation from the effects of space weather, including vulnerability reduction. Mitigation focuses on minimizing risks, addressing cascading effects, and enhancing disaster resilience.[4] Implementation of these preparedness

missions requires joint action from public and private stakeholders whose shared expertise and responsibilities are embedded in the Nation's infrastructure systems.

4) **Improve Assessment, Modeling, and Prediction of Impacts on Critical Infrastructure:** Timely, reliable, actionable, and relevant decision-support services during space-weather events are essential to improving national preparedness. Societal effects must be understood to better inform the actions necessary during extreme events and to encourage appropriate mitigation and protection measures before an incident.

5) **Improve Space-Weather Services through Advancing Understanding and Forecasting:** Opportunity exists to improve the fundamental understanding of space weather and increase the accuracy, reliability, and timeliness of space-weather observations and forecasts (and related products and services). The underpinning science and observations will help drive advances in modeling capability and improve the quality of space-weather products and services. There is also a need to improve capacity to develop and transition the latest scientific and technological advances into space-weather operations centers.

6) **Increase International Cooperation:** In a world of complex interdependencies, global engagement and a coordinated international response to space weather is needed. The United States must not only be an integral part of the global effort to prepare for space-weather impacts, but must also help mobilize broad, global support for this effort by using existing agreements and building international support and policies.

The Strategy identifies goals and establishes the guiding principles that will underpin the Nation's efforts to secure the infrastructures vital to national security and economy of the United States. It identifies specific initiatives to drive both near- and long-term national protection priorities. It also provides protocols for preparing and responding to space-weather events and for ensuring that information is available to inform decision-making. This information will be used to enhance national resilience and prepare an appropriate response during space-weather storms.

This Strategy and the associated Action Plan will facilitate the integration of space-weather information into Federal risk-management plans to achieve preparedness levels consistent with national policies. Accomplishing the strategic elements in the Strategy will require a whole-community approach to coordinating domestic and international public and private resources.[5] Government, industry, and the American people must work together to enhance the resilience of critical infrastructure to the adverse effects of space weather on the people, economy, and security of the Nation.

INTRODUCTION

Space-weather events are naturally occurring phenomena that have the potential to negatively affect technology and energy infrastructure, which are essential contributors to national security and economic vitality. The term "space weather" refers to the dynamic conditions of the space environment that arise from emissions from the sun, which include solar flares, solar energetic particles, and coronal mass ejections (CME).[6] These emissions can interact with Earth and its surrounding space, including the Earth's magnetic field, potentially disrupting electric power systems; satellite, aircraft, and spacecraft operations; telecommunications; position, navigation, and timing services; and other technologies and infrastructures.[7] The Nation's critical infrastructures make up a diverse, complex, interdependent system of systems in which the failure of one could cascade to another. Given the importance of reliable electric power and space-based assets for security and economic well-being, it is essential that the United States establish a strategy to improve the Nation's ability to protect, mitigate, respond to, and recover from the potentially devastating effects of space-weather events.

Space-weather events occur regularly and have measurable effects on critical infrastructure systems and technologies. The *National Space Weather Strategy* (Strategy) and *National Space Weather Action Plan* (Action Plan) establish goals and actions to enhance the understanding of risk from, and national preparedness for, extreme space-weather events. Many of the goals and activities outlined in the Strategy and Action Plan can be scaled to address space-weather events that are smaller in magnitude. Such events occur more frequently than extreme events and can have significant effects.

Space weather is a global issue. Unlike terrestrial weather events (e.g., a hurricane), space weather has the potential to simultaneously affect the whole of North America or reach even wider geographic regions of the planet. Even though the United States is a global leader in observing and forecasting space-weather events, these capabilities depend on international cooperation and coordination.

This Strategy outlines objectives for enhancing the Nation's space-weather readiness in three key areas: national preparedness, forecasting, and understanding. Federal departments and agencies have taken significant steps in these key areas. The challenges posed by global vulnerability to space-weather events require continuing research and development to improve observation and forecasting capabilities, which are linked directly to preparedness.

The goals outlined in this Strategy will leverage these efforts and existing policies, while promoting enhanced coordination and cooperation across the public and private sectors in the United States and abroad.

IMPLEMENTATION OF THE NATIONAL SPACE WEATHER STRATEGY

The Action Plan, released concurrently with this Strategy, details the Federal activities that will be undertaken to implement the Strategy and achieve the six high-level goals, and includes deliverables and timelines. This Strategy acknowledges the challenges associated with planning and preparing for extreme events that do not currently have well-defined recurrence rates; identified activities in the Action Plan should therefore be prioritized accordingly.

The Executive Office of the President will coordinate the execution of the Action Plan and will reevaluate and update the Strategy and Action Plan within 3 years of the date of publication, or as needed.

Full implementation of this Strategy will require the action of a nationwide network of governments, agencies, emergency managers, academia, the media, the insurance industry, nonprofit organizations, and the private sector. Strong public-private collaborations must be established between the Federal Government, industry, and academia to enhance observing networks, conduct research, develop prediction models, and supply the services necessary to protect life and property and to promote economic prosperity. These partnerships will form the backbone of a space-weather-ready Nation.

ENHANCING NATIONAL PREPAREDNESS AND CRITICAL INFRASTRUCTURE RESILIENCE

This Strategy ensures that space weather is fully integrated into the frameworks of two Presidential Policy Directives (PPDs): PPD-8, "National Preparedness" (March 30, 2011); and PPD-21, "Critical Infrastructure Security and Resilience" (February 12, 2013).

PPD-8 calls for an integrated, all-of-Nation, capabilities-based approach to preparedness for all hazards. It also calls for the creation of a series of National Planning Frameworks. Accordingly, the Department of Homeland Security (DHS) coordinated the development of the Strategic National Risk Assessment (SNRA).[8] The SNRA identified space weather as one of nine natural hazards with the potential to significantly affect homeland security.

PPD-21 identifies three strategic imperatives to drive the Federal approach to strengthening critical infrastructure security and resilience at the core of this Strategy.[9] The Directive identifies energy and communications systems as vital due to the enabling functions they provide across all critical infrastructure sectors. The Directive also instructs the Federal Government to engage with industry and international partners to strengthen the security and resilience of domestic and international critical infrastructures on which the Nation depends.

STRATEGIC GOALS

This Strategy defines six strategic goals to prepare the Nation for near- and long-term space-weather impacts. The goals aim to improve the Nation's preparedness for, forecasting of, and understanding of space-weather events, encompassing efforts related both to the phenomena that cause space weather and the effects of these phenomena. (See the Appendix for background on the phenomena that cause space weather.)

The six high-level goals for Federal research, development, deployment, operations, coordination, and engagement are:

1) Establish Benchmarks for Space-Weather Events
2) Enhance Response and Recovery Capabilities
3) Improve Protection and Mitigation Efforts

4) Improve Assessment, Modeling, and Prediction of Impacts on Critical Infrastructure
5) Improve Space-Weather Services through Advancing Understanding and Forecasting
6) Increase International Cooperation.

1. Establish Benchmarks for Space-Weather Events

Benchmarks are a set of characteristics and conditions against which a space-weather event can be measured. They provide a point of reference from which to improve the understanding of space-weather effects, develop more effective mitigation procedures, enhance response and recovery planning, and understand risk.

Benchmarks should provide clear and consistent descriptions of the relevant physical parameters of space-weather phenomena based on current scientific understanding and the historical record. For example, benchmarks may serve as input to vulnerability assessments to help establish decision points and thresholds for action and to inform practices (e.g., device development, operational planning, and mitigation efforts). These benchmarks will not seek to categorize or classify the degree of impact from a space-weather event on a technology or infrastructure system.

To be effective, the benchmarks must be developed in a timely manner using transparent methodology with a clear statement of assumptions and uncertainties and publicly available data (where possible). Because of relatively limited data and gaps in understanding space-weather phenomena, benchmarks should be reevaluated as significant new data and research become available. The following objectives should be pursued in the development of these benchmarks:

- **Define scope, purpose, and approach for developing benchmarks:** The benchmarks will use multiple physical parameters to describe a space-weather event. The parameters should include characteristics of an event and its interactions with Earth and near-Earth environments (e.g., geomagnetic and ionospheric disturbances).
- **Create multiple benchmarks to address different circumstances:** The benchmarks should cover:

- Different types of space-weather events (e.g., ionospheric disturbances induced by solar flares, and geomagnetic disturbances induced by CMEs);
- Multiple physical parameters that will ensure the functionality of the benchmarks (e.g., magnitude and duration); and
- A range of event magnitudes and associated recurrence intervals (e.g., multiple event scenarios may inform different vulnerability thresholds, and an understanding of the worst-case scenario may be instructive).

2. Enhance Response and Recovery Capabilities

Extreme space-weather events are potentially high-impact events that will require a coordinated national response and recovery effort. Leveraging the National Planning Frameworks,[10] the Nation will develop comprehensive guidance to support existing response and recovery capabilities to manage extreme events with government (Federal, State, and local), industry, and other partners. Improved vulnerability assessments and systems modeling will enhance planning for the effects of extreme events on critical infrastructure systems and the Whole Community,[11] as well as inform estimates of duration and costs of response and recovery measures. Likewise, improved forecasting capabilities will enable the development of time-sensitive procedures before significant impacts can occur. Enhancing the Nation's response and recovery capabilities will require continued investments, unique solutions, and strong public-private partnerships. The following objectives should be pursued to enhance response and recovery capabilities:

- **Complete an all-hazards power outage response and recovery plan:** The primary risk from an extreme space-weather event is the potential for the long-term loss of electric power and the cascading effects that it would have on other critical infrastructure sectors. Other high-impact events are also capable of causing long-term regional or national power outages. It is essential to have a comprehensive and executable plan (with key decision points) to address regional or national power outages. The plan must include the Whole Community and prioritize core capabilities.[12]

- **Support government and private-sector planning for and management of extreme space-weather events:** The incorporation of space-weather event information into all-hazards planning is limited for Federal, State, and local governments. Credible information and guidance on how to obtain that knowledge and incorporate it into government all-hazards planning should be developed and disseminated.
- **Provide guidance on contingency planning for the effects of extreme space weather for essential government and industry services:** Preservation of government services, personnel movement, and maintenance of infrastructure systems are essential before, during, and after an extreme space-weather event. Government, the private sector, and critical infrastructure entities need guidance on how to respond in a manner that increases the likelihood of maintaining essential operational elements for a prolonged period of time.
- **Ensure the capability and interoperability of communications systems during extreme space-weather events:** Effective communications systems are essential to gaining and maintaining situational awareness and ensuring unity of effort in response and recovery operations. The effects of space weather on communications systems occur at different timescales and at varying degrees within a single event, depending on the system and the characteristics and duration of the event. Government and private-sector stakeholders need guidance that allows them to maintain communications capabilities (including interoperability) during an extreme space-weather event.
- **Encourage owners and operators of infrastructure and technology assets to coordinate development of realistic power-restoration priorities and expectations:** Electrical power providers should develop protocols for restoring electrical power before disruptions, in coordination with State and local governments. Critical-asset owners and operators must work with their providers to ensure that their power needs are understood. The owners and operators should consider plans and capabilities for temporary power in the event of an electrical power disruption caused by an extreme space-weather event.

- **Develop and conduct exercises to improve and test government and industry-related space-weather response and recovery plans:** Evaluating the effectiveness of plans includes developing and executing a combination of training events and exercises to determine whether the goals, objectives, decisions, actions, and timing outlined in the plans support successful response and recovery. Exercising plans and capturing lessons learned enables ongoing improvement in event response and recovery capabilities.

3. Improve Protection and Mitigation Efforts

Growing interdependencies of critical infrastructure systems have increased potential vulnerabilities to space-weather events. Protection and mitigation efforts to eliminate or reduce space-weather vulnerabilities are essential components of national preparedness. Protection focuses on capabilities and actions to eliminate vulnerabilities to space weather, and mitigation focuses on long-term vulnerability reduction and enhancing resilience to disasters.[13] Together, these preparedness missions frame a national effort to reduce vulnerabilities and manage risks associated with space-weather events. Implementing these preparedness missions requires joint action from both public and private stakeholders, due to the shared expertise and responsibilities embedded in the Nation's infrastructure systems. The following objectives should be pursued to improve protection and mitigation efforts with respect to space-weather events:

- Encourage development of hazard-mitigation plans that reduce vulnerabilities to, manage risks from, and assist with response to the effects of space weather: In support of Whole Community planning for resilience, information about space-weather hazards should be integrated, as appropriate, into existing mechanisms for information sharing, including Information Sharing Analysis Organizations, and into national preparedness mechanisms that promote strategic alignment between public and private sectors.
- Work with industry to achieve long-term reduction of vulnerability to space-weather events by implementing measures at locations most susceptible to space weather: Adopting standards, business practices, and operational procedures that improve protection and resilience is essential to addressing system vulnerabilities to space weather. The

benchmark space-weather events described in the first strategic goal (Establish Benchmarks for Space-Weather Events) should be used to support the adoption of design standards for enhanced resilience; evaluate strategies for, priorities for, and feasibility of protecting critical assets; and foster mechanisms for sharing best practices that promote mitigation and protection of systems affected by space weather.

- Strengthen public-private collaborations that support action to reduce vulnerability to space weather: Private industries are essential to the Nation's resilience. They are the owners and operators of the majority of the Nation's critical infrastructure, and they play a vital role in research and development to enhance understanding and improve mitigation. Space-weather events do not respect national, jurisdictional, or corporate boundaries. Incorporating resilience measures into U.S. infrastructure systems requires public-private collaboration, support of existing coordinating mechanisms for information sharing and access, and identification of incentives and disincentives for investing in resilience measures.

4. Improve Assessment, Modeling, and Prediction of Impacts on Critical Infrastructure

A key component of improving national preparedness for a space-weather event is the ability to observe and predict associated effects. Providing timely, actionable, and relevant decision-support services during a space-weather event requires improvements in abilities to observe, assess, model, and ultimately predict the effects of space-weather events on critical national infrastructures such as electric power systems; transportation systems (e.g., aviation, rail, and maritime); communications; and position, navigation, and timing systems. The societal and health effects of space-weather events must also be understood to inform the urgency of action during such events and to encourage appropriate mitigation and protection measures before an incident. Improving situational awareness and prediction of the effects on infrastructure during a space-weather event requires better observations and better modeling of system-response characteristics. The following objectives should be pursued to enhance observation, modeling, and prediction capabilities:

- **Assess the vulnerability of critical infrastructure systems to space weather:** To prepare for and enhance the security and resilience of critical infrastructure systems to space-weather events, a thorough and systematic understanding of the effects and vulnerabilities is necessary. This understanding will inform preparedness approaches and planning and enable validation of system-specific impact models.

- **Develop a real-time infrastructure assessment and reporting capability:** Situational awareness of the state of various critical infrastructure systems is crucial to providing actionable event response. This capability will require continued investments in, and assessments of, the real-time monitoring requirements for reporting the state of infrastructures, as well as situational awareness of space weather.

- **Develop or refine operational models that forecast the effects of space weather on critical infrastructure:** To ensure an appropriate and effective response to space-weather events, it is not enough to only forecast the magnitude of such events. It is also necessary to predict the effects of such events on infrastructure and other systems on a regional basis. (Hurricane storm-surge prediction is a terrestrial weather example of this objective.) Effective prediction of the effects of space weather requires reliable, accurate, and fast models that take into account effects on both isolated and interdependent infrastructure systems. There is also a need to define and develop comprehensive requirements for operational impact models, identify deficiencies in current modeling capabilities, and develop new and improved tools.

- **Improve operational impact forecasting and communications:** Based on the assessment and modeling elements outlined above, a national capability to forecast extreme space-weather effects before the onset of an event would enable timely warnings to system operators and emergency managers. This capability should always be available, with rapid computation and dissemination mechanisms.

- **Conduct research on the effects of space weather on industries, operational environments, and infrastructure sectors:** Improving existing models and developing new capabilities in impact forecasting must be based on a better understanding of the fundamental physical processes of the effects of space weather on critical infrastructure systems. Doing so requires identifying gaps in understanding of impacts on these systems; developing strategies to address these gaps; identifying impact-related interdependencies through vulnerability and failure mode-assessments across and between sectors; and supporting research for understanding the cost to mitigate, respond to, and recover from extreme space-weather events.

5. Improve Space-Weather Services through Advancing Understanding and Forecasting

Space-weather services can enhance national preparedness by providing timely, accurate, and relevant forecasting products. Identifying and sustaining a baseline of fundamental measurements from observing platforms is key to providing operational services that inform preparedness. This baseline can also serve as a reference point from which to identify coverage and measurement gaps, as well as opportunities for improvement. Services can be improved through basic research and applied research that focus on the needs of an increasingly diverse user community. To facilitate the transition of these enhancements from the research domain to operations, the responsible agencies will: (1) periodically revalidate user requirements for improved space-weather services; and (2) strengthen and encourage partnerships to accelerate the research-to-operations transition process, with a goal to support key preparedness decisions. The following objectives should be pursued to meet these goals:

- **Improve understanding of user needs for space-weather forecasting to establish lead-time and accuracy goals:** Effective transfer of space-weather knowledge requires a better understanding of the effects of space weather on technology and on industry and government customers, including the associated economic and political impacts on the Nation's critical infrastructures.
- **Ensure space-weather products are intelligible and actionable to inform decision-making:** Decision-relevant information must be communicated in ways that stakeholders can fully understand and use. Models and forecasts will be most useful when they enable swift decision-making with a reasonable assumption of risk.
- **Establish and sustain a baseline observational capability for space-weather operations:** The Nation lacks a comprehensive operational space-weather observation strategy. Opportunities exist to improve the Nation's space-weather-prediction capabilities, which rely on an ad hoc mixture of weather satellites, research satellites, and

ground systems to provide data to forecast centers. To ensure adequate and sustained real-time observations for space-weather analysis, forecasting, and decision-support services, a baseline, or minimally adequate, operational observation capability should be defined. The observation baseline should also specify the optimal mix of ground-based and satellite observations to enable continuous and timely space-weather watch, warning, and alert products and services. The associated data reception, relay, processing, assimilation, and archiving infrastructure required to utilize space-weather observations must also be included in the baseline.

- **Improve forecasting lead-time and accuracy:** Society is increasingly at risk from extreme space-weather events. With improved predictions, the Nation can enhance mitigation, response, and recovery actions to safeguard assets and maintain continuity of operations during high-impact space-weather activity.

- **Enhance fundamental understanding of space weather and its drivers to develop and continually improve predictive models:** Forecasting space weather depends on a fundamental understanding of the space-environment processes that give rise to hazardous events. It is particularly important to understand the processes that link the sun to Earth. An improved understanding of space weather and access to better data will help drive the necessary advances in modeling capabilities and validation to support user needs.

- **Improve effectiveness and timeliness of the process that transitions research to operations:** Although the Nation has invested in the development of research infrastructure and predictive models to meet the demands of a growing space-weather user community, existing modeling capabilities still fall short of providing what is needed to meet these demands. Until better research models targeted to operational needs are developed and ultimately incorporated into operational forecasts, the Nation will not fully realize the benefits of its research investments.

6. Increase International Cooperation

In a world increasingly dependent on interconnected and interdependent infrastructure, any disruption to these critical technologies could have regional and even international consequences. Therefore, space weather should be regarded as a global challenge requiring a coordinated global response.

Many countries are becoming increasingly aware of the need to monitor and manage space-weather risks. The United States and other nations are sharing observations and research, disseminating products and services, and collaborating on real-time predictions to mitigate impacts on critical technology and infrastructure. Countries around the world must work together to foster global collaboration, taking advantage of mutual interests and capabilities to improve situational awareness, predictions, and preparedness for extreme space weather. The following objectives should be pursued to increase international cooperation:

- **Build international support and policies for acknowledging space weather as a global challenge:** A prerequisite to enhanced international cooperation is high-level support across partner countries to raise awareness of space weather as a global challenge.
- **Increase engagement with the international community on observation infrastructure, data sharing, numerical modeling, and scientific research:** The Federal Government should explore opportunities to work with the international community to enhance research, observations, models, and forecasting tools that are responsive to the needs of the global scientific community and the providers and users of space-weather information services.
- **Strengthen international coordination and cooperation on space-weather products and services:** Providing high-quality space-weather products and services worldwide requires international consensus and cooperation. Toward this end, the United States should seek international agreement on common terminology, measurements, and scales of magnitude; promote and coordinate sharing and dissemination of space-weather observations, model outputs, and forecasts; and establish coordination procedures across space-weather operations centers during events.
- **Promote a collaborative international approach to preparedness for extreme space-weather events:** The world's interconnected and interdependent systems are vulnerable to extreme space-weather events; this vulnerability could possibly lead to a cascade of impacts across borders and sectors. To mitigate these risks, the United States should work with the international community to

facilitate the exchange of information and best practices to strengthen global preparedness capacity for extreme space-weather events. The United States should also foster the development of global mutual-aid arrangements to facilitate response and recovery efforts, and should coordinate international partnership activities to support space-weather preparedness and response exercises.

CONCLUSION

Space-weather events pose a significant and complex risk to the Nation's infrastructure and have the potential to cause substantial economic and human harm. This Strategy is the first step in addressing the myriad challenges presented when managing and mitigating the risks posed by both extreme and ordinary space weather. The six high-level goals and associated objectives outlined in this Strategy support a collaborative and Federally-coordinated approach to developing effective policies, practices, and procedures for decreasing the Nation's vulnerabilities associated with space weather. By establishing goals for improvements in forecasting, research, preparedness, planning, and domestic and international engagement, this Strategy will help ensure the Nation's resilience to the effects of extreme space-weather events.

APPENDIX: BACKGROUND ON SOLAR PHENOMENA THAT DRIVE SPACE WEATHER

Space weather is primarily driven by solar storm phenomena that include coronal mass ejections (CMEs), solar flares, solar particle events, and solar wind. These phenomena can occur in various regions on the sun's surface, but only Earth-directed solar storms are the potential drivers of space-weather events on Earth. An understanding of solar storm phenomena is an important component to developing accurate space-weather forecasts (event onset, location, duration, and magnitude). CMEs are explosions of plasma (charged particles) from the sun's corona. They generally take 2–3 days to arrive at Earth, but in the most extreme cases they have been observed to arrive in as little as 15 hours. When CMEs collide with Earth's magnetic field, they can cause a space-weather event called a geomagnetic storm, which often includes enhanced auroral displays. Geomagnetic storms of varying magnitudes can cause significant long- and short-term impacts to the Nation's critical infrastructure, including the electric power grid, aviation systems, Global Positioning System (GPS) applications, and satellites.

A solar flare is a brief eruption of intense high-energy electromagnetic radiation from the sun's surface, typically associated with sunspots. Solar flares can affect Earth's upper atmosphere, potentially causing disruption, degradation, or blackout of satellite communications, radar, and high-frequency radio communications. The electromagnetic radiation from the flare takes approximately eight minutes to reach Earth, and the effects usually last for one to three hours on the daylight side of Earth.

Solar particle events are bursts of energetic electrons, protons, alpha particles, and other heavier particles into interplanetary space. Following an event on the sun, the fastest moving particles can reach Earth within tens of minutes and temporarily enhance the radiation level in interplanetary and near-Earth space. When energetic protons collide with satellites or humans in space, they can penetrate deep into the object that they collide with and cause damage to electronic circuits or biological DNA. Solar particle events can also pose a risk to passengers and crew in aircraft at high latitudes near the geomagnetic poles and can make radio communications difficult or nearly impossible.

Solar wind, consisting of plasma, continuously flows from the sun. Different regions of the sun produce winds of different speeds and densities. Solar wind speed and density play an important role in space weather. High-speed winds tend to produce geomagnetic disturbances, and slow-speed winds can bring calm space weather. Space-weather effects on Earth are highly dependent on solar wind speed, solar wind density, and direction of the magnetic field embedded in the solar wind. When high-speed solar wind overtakes slow-speed wind or when the magnetic field of solar wind switches polarity, geomagnetic disturbances can result.

REFERENCES

Blanchard, B. Wayne. "Guide to Emergency Management and Related Terms, Definitions, Concepts, Acronyms, Organizations, Programs, Guidance, Executive Orders & Legislation: A Tutorial on Emergency Management, Broadly Defined, Past and Present." October 22, 2008.

Department of Homeland Security (DHS). *The Strategic National Risk Assessment in Support of PPD 8: A Comprehensive Risk-Based Approach toward a Secure and Resilient Nation.* December 2011.

Federal Emergency Management Agency (FEMA). "National Planning Frameworks." www.fema.gov/national-planning-frameworks.

_____."A Whole Community Approach to Emergency Management: Principles, Themes, and Pathways for Action." FDOC 104-008-1. December 2011, www.fema.gov/media-librarydata/20130726-1813-25045-0649/whole_community_dec2011__2_pdf.

Federal Energy Regulatory Commission (FERC). "Reliability Standards for Geomagnetic Disturbances." Order No. 779, 143 FERC 61,147. May 23, 2013.

_____. "Reliability Standard for Geomagnetic Disturbance Operations." Order No. 797, 147 FERC 61,209. June 25, 2014.

_____.. "Reliability Standard for Transmission System Planned Performance for Geomagnetic Disturbance Events." 151 FERC 61,134 May 14, 2015.

National Research Council. *Severe Space Weather Events—Understanding Societal and Economic Impacts, A Workshop Report*. Committee on the Societal and Economic Impacts of Severe Space Weather Events: A Workshop. 2008.

_____.*Solar and Space Physics: A Science for a Technological Society*. 2013–2022 Decadal Survey in Solar and Space Physics. 2013.

National Science and Technology Council. *National Strategy for Civil Earth Observations*. April 2013. National Space Policy of the United States of America. June 28, 2010.

National Space Weather Program Council. *National Space Weather Program: Strategic Plan*. FCM-P30- 2010, 2010.

Office of the Federal Coordinator for Meteorological Services and Supporting Research. *Report of the Assessment Committee for the National Space Weather Program*. FCM-R24-2006, June 2006.

Presidential Policy Directive/PPD-8. "National Preparedness." March 30, 2011.

Presidential Policy Directive/PPD-21. "Critical Infrastructure Security and Resilience." February 12, 2013.

Public Law 111-267. *The National Aeronautics and Space Administration Authorization Act of 2010*. October 11, 2010.

Schrijver, Carolus J., Kirsti Kauristie, Alan D. Aylward, Clezio M. Denardini,Sarah E. Gibson, Alexi Glover, Nat Gopalswamy, Manuel Grande, Mike Hapgood, Daniel Heynderickx, Norbert Jakowski, Vladimir V. Kalegaev,Giovanni Lapenta, Jon A. Linker, Siqing Liu, Cristina H. Mandrini, Ian R. Mann,Tsutomu Nagatsuma, Dibyendu Nandi, Takahiro Obara, T. Paul O'Brien,Terrance Onsager, Hermann J. Opgenoorth,

Michael Terkildsen, Cesar E. Valladares, and Nicole Vilmer. "Understanding Space Weather to Shield Society: A Global Road Map for 2015–2025 Commissioned by COSPAR and ILWS" *Advances in Space Research* 55 (2015): 2745–2807.

U.S. Department of Energy, Office of Electricity Delivery and Energy Reliability Infrastructure Security and Energy Restoration. *Insurance as a Risk Management Instrument for Energy Infrastructure Security and Resilience*. 2013.

ABOUT THE NATIONAL SCIENCE AND TECHNOLOGY COUNCIL

The National Science and Technology Council (NSTC) is the principal means by which the Executive Branch coordinates science and technology policy across the diverse entities that make up the Federal research and development enterprise. One of the NSTC's primary objectives is establishing clear national goals for Federal science and technology investments. The NSTC prepares R&D packages aimed at accomplishing multiple national goals. The NSTC's work is organized under five committees: Environment, Natural Resources, and Sustainability (CENRS); Homeland and National Security; Science, Technology, Engineering, and Mathematics (STEM) Education; Science; and Technology. Each of these committees oversees subcommittees and working groups that are focused on different aspects of science and technology. More information is available at www.whitehouse.gov/ostp/nstc.

ABOUT THE OFFICE OF SCIENCE AND TECHNOLOGY POLICY

The Office of Science and Technology Policy (OSTP) was established by the National Science and Technology Policy, Organization, and Priorities Act of 1976. OSTP's responsibilities include advising the President in policy formulation and budget development on questions in which science and technology are important elements; articulating the President's science and technology policy and programs; and fostering strong partnerships among Federal, State, and local governments, and the scientific communities in industry and academia. The Director of OSTP also serves as Assistant to the President for Science and Technology and manages the NSTC. More information is available at www.whitehouse.gov/ostp.

ABOUT THE SPACE WEATHER OPERATIONS, RESEARCH, AND MITIGATION (SWORM) TASK FORCE

The Space Weather Operations, Research, and Mitigation (SWORM) task force, an interagency group organized under the NSTC, CENRS, Subcommittee on Disaster Reduction (SDR), was chartered in November 2014 to develop a national strategy and a national action plan to enhance national preparedness for space-weather events.

ABOUT THIS DOCUMENT

This document was developed by the SWORM Task Force. It was released in draft for public comment on the Federal Register (80 FR 24296), was reviewed by SDR and CENRS, and was finalized and published by OSTP.

ACKNOWLEDGMENTS

The SWORM Task Force acknowledges the contributions from the IDA Science and Technology Policy Institute for providing subject-matter expertise, constructive review, and other contributions to the development of this strategy.

SPACE WEATHER OPERATIONS, RESEARCH, AND MITIGATION TASK FORCE

Co-Chairs

Department of Commerce, National Oceanic and Atmospheric Administration Department of Homeland Security Office of Science and Technology Policy

Members

Departments
Department of Commerce Department of Defense
Department of Energy
Department of Homeland Security
Department of the Interior Department of State
Department of Transportation

Agencies and Service Branches
Federal Aviation Administration
Federal Communications Commission
Federal Emergency Management Agency
Federal Energy Regulatory Commission
National Aeronautics and Space Administration
National Oceanic and Atmospheric Administration
National Science Foundation
Nuclear Regulatory Commission
Office of the Director of National Intelligence
United States Air Force
United States Geological Survey
United States Navy
United States Postal Service

Executive Office of the President
National Security Council
Office of Management and Budget
Office of Science and Technology Policy
White House Military Office

End Notes

[1] Department of Homeland Security, The Strategic National Risk Assessment (SNRA) in Support of PPD 8: A Comprehensive Risk-Based Approach toward a Secure and Resilient Nation, December 2011.

[2] See References section for a list of recent relevant policy documents.

[3] Local governments include tribal, territorial, and insular area governments.

[4] Disaster resilience refers to the capability to prevent, or protect infrastructure from, significant multi-hazard threats and incidents and to expeditiously recover and reconstitute critical

services with minimum damage to public safety and health, the economy, and national security.

[5] Whole-community planning for resilience is an approach to emergency management that reinforces the ideas that the Federal Emergency Management Agency (FEMA) is only one part of the Nation's emergency management team; that collective resources must be leveraged in preparing for, protecting against, responding to, recovering from, and mitigating against all hazards; and a collective effort is required to meet the needs of the entire community in each of these areas (FEMA, A Whole Community Approach to Emergency Management: Principles, Themes, and Pathways for Action, FDOC 104- 008-1, December 2011).

[6] Not all effects of space weather are damaging. The aurora borealis is a striking visual manifestation of space weather.

[7] For a more detailed description of space weather and its drivers, please refer to the Appendix.

[8] DHS, The Strategic National Risk Assessment in Support of PPD 8.

[9] (1) Refine and clarify functional relationships across the Federal Government to advance the national unity of effort to strengthen critical infrastructure security and resilience; (2) Enable effective information exchange by identifying baseline data and systems requirements for the Federal Government; and (3) Implement an integration and analysis function to inform planning and operations decisions regarding critical infrastructure.

[10] The National Planning Frameworks describe how the Whole Community works together to achieve the national preparedness goal of "a secure and resilient nation with the capabilities required across the whole community to prevent, protect against, mitigate, respond to, and recover from the threats and hazards that pose the greatest risk." This goal is the cornerstone for the implementation of PPD-8 (FEMA website, "National Planning Frameworks," www.fema.gov/nationalplanning-frameworks).

[11] Whole Community partners refer to the Nation's larger collective emergency management team and include not only DHS and its partners at the Federal level, but also State, local, tribal, and territorial (SLTT) partners, non-governmental organizations such as faith-based and nonprofit groups and private sector industry, and individuals, families and communities. See FEMA website, "Whole Community," last updated April 16, 2015, www.fema.gov/whole-community.

[12] The FEMA National Response Framework (May 2013) defines 14 response-centric core capabilities.

[13] Disaster resilience refers to the capability to prevent, or protect infrastructure from, significant multi-hazard threats and incidents and to expeditiously recover and reconstitute critical services with minimum damage to public safety and health, the economy, and national security (Blanchard, "Guide to Emergency Management and Related Terms").

In: Space Weather
Editor: Peter Burton

ISBN: 978-1-63484-440-6
© 2016 Nova Science Publishers, Inc.

Chapter 2

NATIONAL SPACE WEATHER ACTION PLAN[*]

Space Weather Operations, Research, and Mitigation Task Force

INTRODUCTION

Space-weather events are naturally occurring phenomena that have the potential to disrupt electric power systems; satellite, aircraft, and spacecraft operations; telecommunications; position, navigation, and timing services; and other technologies and infrastructures that contribute to the Nation's security and economic vitality. These critical infrastructures make up a diverse, complex, interdependent system of systems in which a failure of one could cascade to another. Given the importance of reliable electric power and space-based assets, it is essential that the United States has the ability to protect, mitigate, respond to, and recover from the potentially devastating effects of space weather.

The *National Space Weather Strategy* (Strategy), released concurrently with this *National Space Weather Action Plan* (Action Plan), details national goals for leveraging existing policies and ongoing research and development efforts regarding space weather while promoting enhanced domestic and international coordination and cooperation across public and private sectors. The implementation of the Strategy will require the action of a nationwide

---

[*] This is an edited, reformatted and augmented version of a document issued by the Office of Science and Technology Policy, National Science and Technology Council, October 2015.

network of governments, agencies, emergency managers, academia, the media, the insurance industry, nonprofit organizations, and the private sector. Strong public-private partnerships must be established to enhance observing networks, conduct research, develop prediction models, and supply the services necessary to protect life and property and to promote economic prosperity. These partnerships will form the backbone of a space-weather-ready Nation. This Action Plan details the activities, outcomes, and timelines that will be undertaken by Federal departments and agencies for the Nation to make progress toward the Strategy's goals.

Structure of the Action Plan

With the objectives of improving understanding of, forecasting of, and preparedness for space-weather events (both the phenomena and their effects), the *National Space Weather Strategy* defines six strategic goals to prepare the Nation for near- and long-term space-weather effects. This Action Plan is organized around the same six strategic goals:

1) Establish Benchmarks for Space-Weather Events
2) Enhance Response and Recovery Capabilities
3) Improve Protection and Mitigation Efforts
4) Improve Assessment, Modeling, and Prediction of Impacts on Critical Infrastructure
5) Improve Space-Weather Services through Advancing Understanding and Forecasting
6) Increase International Cooperation

Implementation of the National Space Weather Strategy

The Action Plan outlines the Federal implementation approach for the *National Space Weather Strategy*. Each action has an associated deliverable and timeline. An interagency body coordinated by the Executive Office of the President will help oversee these actions, including the deliverables and timelines. This Action Plan aligns with investments proposed in the President's Budget for Fiscal Year 2016. The interagency body will also reevaluate and update the Strategy and Action Plan within 3 years of the date of publication, or as needed. Successful progress in achieving these objectives depends on a national commitment of resources, both public and private.

Each action indicates the lead Federal Executive Branch department for completing the action, but does not prescribe a specific approach. For example, a department might engage academia or the private sector to complete a study or procure data. This Action Plan acknowledges the challenges associated with planning and preparing for extreme events that do not currently have well-defined recurrence rates; identified activities should therefore be prioritized accordingly, consistent with existing authorities.

> Note: The actions specified in the *National Space Weather Action Plan* are intended to inform the policy development process and are not intended as a budget document. The commitment of Federal resources to support these activities will be determined through the budget process. Additional resources may be needed to fully implement many of the actions in this report; such resources could come from new Federal appropriations, redirection from lower priority Federal Agency activities, and/or from State, local, and/or other resources.

GOAL 1: ESTABLISH BENCHMARKS FOR SPACE-WEATHER EVENTS

Introduction

Benchmarks are a set of physical characteristics and conditions against which a space-weather event can be measured. They describe the nature and intensity of extreme space-weather events, providing a point of reference from which to improve understanding of space-weather effects. Benchmarks can serve as input for creating engineering standards, developing vulnerability assessments, establishing decision points and thresholds for action, understanding risk, developing more-effective mitigation procedures and practices, and enhancing response and recovery planning.

Space-weather benchmarks should provide clear and consistent descriptions of space-weather events based on current scientific understanding and the historical record. These benchmarks will not seek to categorize or classify a space-weather event's degree of impact on a technology system.

A two-phase approach with different timelines will be used to balance immediate needs with requirements for scientifically and statistically rigorous benchmarks. Phase 1, a quick-turnaround analysis, will seek to develop each benchmark using existing data sets and studies, where available. For those benchmarks where a quick-turnaround analysis will not yield results of sufficient quality, more rigorous analyses (Phase 2) will be conducted. Benchmarks that are successful in Phase 1 may benefit from further refinement in Phase 2. All benchmarks will be reexamined at least every 5 years or when significant new data or models become available.

The objectives of Goal 1 are to develop the following benchmarks:

- Induced geo-electric fields
- Ionizing radiation
- Ionospheric disturbances
- Solar radio bursts
- Upper atmospheric expansion

1.1. Develop Benchmarks for Induced Geo-Electric Fields

Geomagnetic storms can induce geo-electric fields in the Earth's crust, driving electric currents in long conductors on or near the Earth's surface. These induced geo-electric fields present a risk to the reliable operation of electric power systems and may affect gas and oil pipelines, railways, and other infrastructures that have long conductive paths. For example, a geo-electric field induced by a space-weather event can produce electric currents (i.e., geomagnetically induced currents [GICs]) that could affect electric-grid system stability, with the potential to damage or cause the failure of essential electric power transmission components. Depending on the severity of the geomagnetic storm, cascading system failure or damage could lead to regional interruptions of electrical power distribution and result in complications with recovery and restoration efforts. To be useful, geo-electric field benchmarks should characterize the induced geo-electric field at the Earth's surface (E-field). This parameter can feed directly into vulnerability studies conducted by industry and the private sector.

At a minimum, the E-field benchmarks and associated confidence levels will define the following:

- Amplitude of the induced E-field; and
- Time dependence of the induced E-field.

At a minimum, these benchmarks will be developed for the following event occurrence rate and intensity level:

- An occurrence frequency of 1 in 100 years; and
- An intensity level at the theoretical maximum for the event.

All benchmarks will state the assumptions made and the associated uncertainties and provide sufficient means to account for regional differences across the United States.

The following actions will be taken to develop induced geo-electric field benchmarks:

1.1.1. The Department of the Interior (DOI), the Department of Commerce (DOC), and the National Aeronautics and Space Administration (NASA), in coordination with the Department of Homeland Security (DHS), the Department of Energy (DOE), and the National Science Foundation (NSF), will: (1) assess the feasibility of establishing functional benchmarks using currently available storm data sets, existing models, and published literature; and (2) use the existing body of work to produce benchmarks for specific regions of the United States.
    Deliverable: Develop Phase 1 benchmarks
    Timeline: Within 180 days of the publication of this Action Plan

1.1.2. DOI, DOC, NASA, and NSF, in coordination with DHS and DOE, will assess the suitability of current data sets and methods to develop a more-refined (compared to Phase 1) set of benchmarks. The assessment will also identify gaps in methods and available data, project the cost of filling these gaps, and project the potential improvement to the benchmarks based on filling each gap.
    Deliverable: Complete assessment report
    Timeline: Within 1 year of the publication of this Action Plan

1.1.3. DOI, DOC, NASA, and NSF, in coordination with DHS and DOE, will improve on the induced geo-electric field benchmarks for the continental United States.

Deliverable: Develop (Phase 2) improved induced geo-electric field benchmarks

Timeline: Within 2 years of the publication of this Action Plan

1.2. Develop Benchmarks for Ionizing Radiation

Changes in the near-Earth radiation environment can affect satellite operations, astronauts in space, commercial space activities, and the radiation environment on aircraft at relevant latitudes or altitudes. Understanding the diverse effects of increased radiation is challenging, but the ionizing radiation benchmarks will help address these effects.

The following areas should be considered in addressing the near-Earth radiation environment:[1] the Earth's trapped radiation belts, the galactic cosmic ray background, and solar energetic-particle events. The radiation benchmarks should account for any change in the near-Earth radiation environment, which, under extreme cases, could present a significant risk to critical infrastructure operations or human health.

At a minimum, the ionizing radiation benchmarks and associated confidence levels will define at least the radiation intensity as a function of time, particle type, and energy for the following event-occurrence rate and intensity level:

- An occurrence frequency of 1 in 100 years; and
- An intensity level at the theoretical maximum for the event.

The benchmarks will address radiation levels at all applicable altitudes and latitudes in the near-Earth environment, and all benchmarks will state the assumptions made and the associated uncertainties.

The following actions will be taken to develop ionizing radiation benchmarks:

1.2.1. NASA and DOC, in coordination with NSF, the Department of Transportation (DOT), the Department of Defense (DOD), and the Federal Communications Commission (FCC), will: (1) assess the feasibility and utility of establishing functional benchmarks for ionizing radiation using the existing models and body of literature for this phenomenon; and (2) use the existing body of work to produce benchmarks.

Deliverable: Develop Phase 1 benchmarks

Timeline: Within 180 days of the publication of this Action Plan

1.2.2. NASA and DOC, in coordination with NSF, DOT, DOD, and FCC, will assess the suitability of current data sets and methods to develop a more-refined (compared to Phase 1) set of benchmarks. The assessment will identify gaps in methods and available data, project the cost of filling the gaps, and project the improvement to the benchmarks based on filling each gap.
  Deliverable: Complete assessment report
  Timeline: Within 1 year of the publication of this Action Plan

1.2.3. NASA and DOC, in coordination with NSF, DOT, DOD, and FCC, will develop enhanced benchmarks.
  Deliverable: Develop (Phase 2) improved ionizing radiation benchmarks
  Timeline: Within 2 years of the publication of this Action Plan

1.3. Develop Benchmarks for Ionospheric Disturbances

Ionospheric disturbances can adversely affect radio signals that propagate through the upper atmosphere, disrupting communication, navigation, and surveillance capabilities over wide areas on timescales ranging from minutes to hours.[2] These disturbances can be caused directly by solar flares or indirectly by interactions between the solar wind and the Earth's magnetic field. Solar flares primarily affect the dayside of the Earth, while solar wind can affect both the dayside and nightside.

High-frequency radio signals, which are used for airline, maritime, and emergency communications, are particularly susceptible to ionospheric disturbances.[3] Effects of ionospheric disturbances can limit or restrict polar-route flights by commercial and military aircraft for several days and block amateur radio communications that are often used as a backup in disaster situations. Ionospheric disturbances induced by space-weather events can also produce signal errors in position, navigation, and timing (PNT) systems such as the Global Positioning System (GPS).

At a minimum, the ionospheric disturbance benchmarks and associated confidence levels will define at least the following:

- Ionospheric radio absorption and duration as a function of frequency;
- Total electron content (slant, vertical, and rate of change);
- Ionospheric refractive index; and

- Peak ionospheric densities and the height of the peak.

At a minimum, these benchmarks will be developed for at least the following event-occurrence rate and intensity level:

- An occurrence frequency of 1 in 100 years; and
- An intensity level at the theoretical maximum for the event.

All benchmarks will state the assumptions made and the associated uncertainties. The following actions will be taken to develop ionospheric disturbance benchmarks:

1.3.1. DOC and DOD, in coordination with NASA, DOI, NSF, and FCC, will: (1) assess the feasibility and utility of establishing functional benchmarks using the existing models and body of literature for this phenomenon; and (2) use the existing body of work to produce benchmarks.
   Deliverable: Develop Phase 1 benchmarks
   Timeline: Within 180 days of the publication of this Action Plan

1.3.2. DOC and DOD, in coordination with NASA, DOI, NSF, and FCC, will assess the suitability of current data sets and methods to develop a more-refined (compared to Phase 1) set of benchmarks. The assessment will identify gaps in methods and available data, project the cost of filling the gaps, and project the improvement to the benchmarks based on filling each gap.
   Deliverable: Complete assessment report
   Timeline: Within 1 year of the publication of this Action Plan

1.3.3. DOC and DOD, in coordination with NASA, DOI, NSF, and FCC, will develop enhanced benchmarks.
   Deliverable: Develop (Phase 2) improved ionospheric disturbance benchmarks
   Timeline: Within 2 years of the publication of this Action Plan

1.4. Develop Benchmarks for Solar Radio Bursts (SRBs)

SRBs are radio wave emissions from the sun that can interfere with radar, communication, and tracking signals. In severe cases, SRBs can inhibit the successful use of radio communications and disrupt a wide range of systems that are reliant on PNT services on timescales ranging from minutes to hours across wide areas on the dayside of Earth.

Solar flares are the primary drivers of SRBs. Intense SRBs interfere with radar systems, satellite communications, and PNT by increasing the noise level at frequencies of operation and overloading the signal, thus making it difficult for ground facilities to operate. Different types of radio bursts affect different frequency ranges.

At a minimum, the SRB benchmarks and associated confidence levels will define at least the following:

- Wavelength or frequency bands of the relevant SRBs; and
- Flux (photon energy per unit area) in these bands (solar flux unit).

At a minimum, these benchmarks will be developed for at least the following event-occurrence rate and intensity level:

- An occurrence frequency of 1 in 100 years; and
- An intensity level at the theoretical maximum for the event.

All benchmarks will state the assumptions made and the associated uncertainties. The following actions will be taken to develop SRB benchmarks:

1.4.1. DOC, DOD, and NASA, in coordination with DOI and FCC, will: (1) assess the feasibility and utility of establishing functional benchmarks using the existing models and body of literature for this phenomenon; and (2) use the existing body of work to produce benchmarks.
Deliverable: Develop Phase 1 benchmarks
Timeline: Within 180 days of the publication of this Action Plan

1.4.2. DOC, DOD, and NASA, in coordination with DOI and FCC, will assess the suitability of current data sets and methods to develop a more-refined (compared to Phase 1) set of benchmarks. The assessment will identify gaps in methods and available data, project the cost of filling the gaps, and project the improvement to the benchmarks based on filling each gap.
Deliverable: Complete assessment report
Timeline: Within 1 year of the publication of this Action Plan

1.4.3. DOC, DOD, and NASA, in coordination with DOI and FCC, will develop enhanced benchmarks.

Deliverable: Develop (Phase 2) improved SRB benchmarks
Timeline: Within 2 years of the publication of this Action Plan

1.5. Develop Benchmarks for Upper Atmospheric Expansion

Space-based systems support many different critical infrastructure sectors necessary to ensure the Nation's homeland and economic security. Many of these systems are located in low-Earth orbit (LEO). Upper-atmospheric expansion describes an increase in the temperature and density of the Earth's upper atmosphere. This change is driven by solar activity and can have a direct impact on LEO spacecraft that are susceptible to the effects of atmospheric drag.[4] Increased drag can pull satellites closer to Earth, changing their orbit, decreasing the lifespan of space assets, and making satellite tracking difficult. Understanding this phenomenon and its relationship to atmospheric drag is essential to maintain safe and effective operations of space-based assets through collision avoidance, accurate satellite tracking, object custody, and reentry prediction. At a minimum, the upper-atmospheric expansion benchmarks and associated confidence levels will define neutral density, winds, composition, and temperature of the thermosphere for the following event-occurrence rate and intensity level:

- An occurrence frequency of 1 in 100 years; and
- An intensity level at the theoretical maximum for effects.

All benchmarks will state the assumptions made and the associated uncertainties. The following actions will be taken to develop upper-atmospheric expansion benchmarks:

1.5.1. DOC, DOD, NSF, and NASA, in coordination with DOI and FCC, will: (1) assess the feasibility and utility of establishing functional benchmarks using the existing models and body of literature for this phenomenon; and (2) use the existing body of work to produce benchmarks.

Deliverable: Develop and complete Phase 1 benchmarks
Timeline: Within 180 days of the publication of this Action Plan

1.5.2. DOC, DOD, NSF, and NASA, in coordination with DOI and FCC, will assess the suitability of current data sets and methods to develop a more-refined (compared to Phase 1) set of benchmarks. The assessment will identify gaps in methods and available data, project the cost of filling the gaps, and project the improvement to the benchmarks based on filling each gap.

Deliverable: Complete assessment report

Timeline: Within 1 year of the publication of this Action Plan

1.5.3. DOC, DOD, NSF, and NASA, in coordination with DOI and FCC, will develop enhanced benchmarks.

Deliverable: Develop (Phase 2) improved upper atmospheric expansion benchmarks

Timeline: Within 2 years of the publication of this Action Plan

GOAL 2: ENHANCE RESPONSE AND RECOVERY CAPABILITIES

Introduction

The effects of space-weather events on critical infrastructure systems and economic sectors depend on the severity of the event. The effects of an extreme space-weather event may require a coordinated national response and recovery.

The Nation can leverage the plans and frameworks in the National Planning Frameworks[5] to help manage response and recovery. From these frameworks, the Nation will develop comprehensive guidance to support and improve existing response and recovery capabilities with Federal, State, and local government,[6] and other Whole Community[7] partners.

An improved ability to forecast and understand the effects and magnitude of a space-weather event is key to enhancing response and recovery planning. Actions that seek to enhance forecasting, impact assessment, and benchmarking of events are outlined in Goals 5, 4, and 1 (respectively) of this Action Plan. Building the Nation's restoration capability will require continued investments, unique solutions, and strong public-private collaborations.

The objectives for Goal 2 are:

- Complete an all-hazards power outage response and recovery plan
- Support government and private-sector planning for and management of extreme space-weather events
- Provide guidance on contingency planning for the effects of extreme space weather for essential government and industry services
- Ensure the capability and interoperability of communications systems during extreme space-weather events
- Encourage owners and operators of infrastructure and technology assets to coordinate development of realistic power-restoration priorities and expectations
- Develop and conduct exercises to improve and test government and industry-related space-weather response and recovery plans

2.1. Complete an All-Hazards Power Outage Response and Recovery Plan

2.1.1. DHS, in partnership with DOE, will develop an all-hazard Power Outage Incident Annex (POIA) to the Federal Interagency Operations Plans (FIOPs)[8] for response and recovery that includes the response to and recovery from an extreme space-weather event.

Deliverable: Complete POIA
Timeline: Within 120 days of the publication of this Action Plan

2.2. Support Government and Private-Sector Planning for and Management of Extreme Space-Weather Events

2.2.1. DHS, in coordination with NASA and DOC, will incorporate the latest data on the threats and vulnerabilities from extreme space weather into the next Strategic National Risk Assessment (SNRA).

Deliverable: Complete update to SNRA
Timeline: Within 2 years of the publication of this Action Plan and update every 3 years thereafter

2.2.2. DHS, in coordination with DOC, DOD, and NASA, will ensure a consistent, joint message concerning the research, prediction, and preparedness for extreme space-weather events across the Federal Government.

Deliverable: Complete development of linkages between agency websites and ensure consistency in messaging

Timeline: Within 120 days of the publication of this Action Plan

2.2.3. DHS, in coordination with the National Preparedness Goal activity,[9] will set forth procedures for accessing and using available space-weather forecast and impact assessment modeling tools to inform response and recovery operational decision-making.

Deliverable: Available tools are identified and socialized with the Emergency Support Function Leadership Group (ESFLG)/Recovery Support Function Leadership Group (RSFLG); and referenced in the FEMA National Watch Center procedures

Timeline: Within 210 days of the completion of Phase 1 benchmarks and Action 4.1.1 initial assessments

2.2.5. DOC and DHS will ensure that space-weather products are integrated into established national preparedness plans (e.g., FEMA Recovery FIOP[10]), to including a background on space-weather phenomena, and products and services.

Deliverable: Complete integration of space-weather products and develop a process to integrate future products into established response and recovery plans

Timeline: To be commensurate with national preparedness frameworks and FIOPs update cycle

2.2.6. DHS, through Emergency Support Function (ESF) 15[11] and in coordination with DOC, will develop a template for issuing a public information alert and a template for warning messaging for an impending and ongoing threat of extreme space weather to critical infrastructure, the private sector, State, local, tribal, and territorial (SLTT) governments, communities, and individuals and families.[12]

Deliverable: Complete update to the ESF 15 standard operating procedures

Timeline: Within 1 year of the publication of this Action Plan

2.3. Provide Guidance on Contingency Planning for the Effects of Extreme Space Weather for Essential Government and Industry Services

2.3.1. DHS, in coordination with Sector-Specific Agencies (SSAs),[13] will determine the immediate and cascading impacts of a benchmarked space-weather event on essential government and industry services, and provide guidelines on the inclusion of these impacts for continuity and contingency for all-hazards planning and exercises. Guidance will include the application of established benchmarks (as described in Goal 1), making operational the available forecast and impact assessment models or tools (as described in Goals 4 and 5), integration with protection and mitigation efforts (as described in Goal 3), and collaboration with spaceweather-community partners. Specific attention to pre-event warning and protective measures will be included in continuity planning guidance.

Deliverable: Complete a National Risk Estimate for Space Weather[14] and integrate comprehensive space-weather preparedness into existing all-hazard preparedness guidance

Timeline: Within 1 year of the completion of Phase 1 benchmarks and Action 4.1.1 initial assessments

2.4. Ensure the Capability and Interoperability of Communications Systems during Extreme Space-Weather Events

2.4.1. DHS will assess the dependencies and vulnerabilities of the various communications systems used by government and industry to support response and recovery operations in the wake of an extreme space-weather event.

Deliverable: Complete white paper on space-weather communications assessment

Timeline: Within 120 days of the completion of Action 2.3.1

2.4.2. DHS will develop guidance, including planning factors, on operating communications systems during and after a benchmarked space-weather event.

Deliverable: Complete comprehensive communications systems operations guidance

Timeline: Within 120 days of the completion of Action 2.4.1

2.5. Encourage Owners and Operators of Infrastructure and Technology Assets to Coordinate Development of Realistic Power-Restoration Priorities and Expectations

2.5.1. DHS, in coordination with DOD, will identify which essential facilities have sufficient back-up power capability to survive an extreme space-weather event and which have the ability to quickly deploy or accept temporary power.

Deliverable: Enter temporary power data into Emergency Power Facility Assessment Tool (EPFAT) database

Timeline: Ongoing

2.6. Develop and Conduct Exercises to Improve and Test Government and Industry-Related Space-Weather Response and Recovery Plans

2.6.1 DHS, in coordination with DOC, DOD, NASA, and DOT, will develop training materials to familiarize scientific, national security, and emergency management professionals with the role and execution of emergency management protocols during the response to extreme space-weather events.

Deliverable: Produce training materials and conduct an annual training seminar through an existing coordination forum

Timeline: Within 1 year of the completion of Phase 1 benchmarks and Action 5.2.2

2.6.2. DHS will incorporate exercise objectives tailored to testing and evaluating the Nation's capabilities to respond to and recover from the potential impacts of a benchmarked space-weather event within relevant exercises.

Deliverable: Incorporate exercise objectives appropriate to space weather in exercise plans

Timeline: Within 180 days of the completion of Action 2.1.1

GOAL 3: IMPROVE PROTECTION AND MITIGATION EFFORTS

Introduction

Growing interdependencies among critical infrastructure systems and increasing reliance on electronic technologies have increased the Nation's vulnerability to space-weather events. Protection and mitigation efforts to eliminate or reduce space-weather vulnerabilities are essential missions of national preparedness. Protection focuses on developing capabilities and actions to secure the Nation from the effects of space weather, including vulnerability reduction. Mitigation focuses on minimizing risks, addressing cascading effects, and enhancing the resilience to disasters.[15] Together, these preparedness missions frame a national effort to reduce the vulnerabilities and manage the risks associated with space-weather events. Implementation of these missions requires joint action from public and private stakeholders, due to the shared expertise and responsibilities embedded in the Nation's infrastructure systems.

The objectives for Goal 3 are:

- Encourage development of hazard-mitigation plans that reduce vulnerabilities to, manage risks from, and assist with response to the effects of space weather
- Work with industry to achieve long-term reduction of vulnerability to space-weather events by implementing measures at locations most susceptible to space weather
- Strengthen public-private collaborations that support action to reduce vulnerability to space weather

3.1. Encourage Development of Hazard-Mitigation Plans that Reduce Vulnerabilities to, Manage Risks from, and Assist with Response to the Effects of Space Weather

3.1.1. DHS, in support of Whole Community planning for resilience, will integrate information about space-weather hazards into existing mechanisms for information sharing, including Sector Coordinating Councils (SCCs), and into national preparedness mechanisms that promote strategic alignment between public and private sectors.

Deliverable: Complete update to relevant planning documents
Timeline: Within 1 year of the publication of this Action Plan

3.1.2. DHS will develop a guidance document for integrating space-weather mitigation into existing coordinating mechanisms for mitigation and protection, including Federal leadership groups and SCCs, as outlined in the National Infrastructure Protection Plan.[16]

Deliverable: Complete guidance document
Timeline: Within 1 year of the publication of this Action Plan

3.1.3. DHS, in coordination with other agencies as appropriate, will provide protection and mitigation guidance to enhance resilience across multiple sectors vulnerable to a range of space-weather events, including events less intense than those specified in the benchmarks (Goal 1). Guidance should be applicable to governments and private-sector owners and operators.

Deliverable: Complete guidance development
Timeline: Within 1 year of the completion of Action 4.1.1 initial assessments

3.2. Work with Industry to Achieve Long-Term Reduction of Vulnerability to Space-Weather Events by Implementing Measures at Locations Most Susceptible to Space Weather

3.2.1. DHS will enhance current and forthcoming strategic national risk assessments, analytic projects, national risk estimates, and other relevant activities that provide vulnerability assessments and prioritization guidance for infrastructure sectors at risk from space weather.

Deliverable: Complete necessary modification of assessments and other relevant activities
Timeline: Within 1 year of the publication of this Action Plan

3.2.2. DHS, in coordination with relevant Federal agencies, will develop Federal resilience guidance for space weather, including tools for assessing the value of backup, redundant, and replacement systems. Based on benchmark development and vulnerability analysis, this guidance will identify which facilities and systems need protection.

Deliverable: Complete resilience guidance
Timeline: Within 2 years of the completion of the Phase 1 benchmarks and Action 4.1.1 initial assessments

3.3. Strengthen Public-Private Collaborations that Support Action to Reduce Vulnerability to Space Weather

3.3.1. DHS, in coordination with relevant Federal agencies, will develop a cross-sector engagement strategy and assess the landscape and feasibility of incentives.
Deliverable: Complete strategy and assessment report
Timeline: Within 1 year of the publication of this Action Plan

GOAL 4: IMPROVE ASSESSMENT, MODELING, AND PREDICTION OF IMPACTS ON CRITICAL INFRASTRUCTURE

Introduction

Many of the fundamental physical characteristics of how space weather affects critical infrastructure systems, such as the electric power system, are not fully understood. To enhance preparedness and inform mitigation, protection, response, and recovery activities, the United States must address these gaps in scientific and engineering understanding. Understanding the effects of space weather and associated infrastructure vulnerabilities will support the creation of an operational forecasting capability[17] of space-weather effects. These capabilities will help enable timely warnings to system operators, policy makers, and emergency managers.

Goal 4 seeks to understand vulnerabilities, increase situational awareness, and develop the capability to predict impacts on all affected critical infrastructure systems. Three infrastructures and technologies of particular concern are the electric power grid, conventional aviation and space travel,[18] and position, navigation, and timing (PNT) systems. Real-time monitoring of effects on critical infrastructure is essential for situational awareness, enhanced preparedness, and model validation. Identifying necessary measurements and ways to enhance data sharing are important activities in enabling the creation of a successful monitoring initiative.

The objectives for Goal 4 are:

- Assess the vulnerability of critical infrastructure systems to space weather
- Develop a real-time infrastructure assessment and reporting capability
- Develop or refine operational models that forecast the effects of space weather on critical infrastructure
- Improve operational impact forecasting and communications
- Conduct research on the effects of space weather on industries, operational environments, and infrastructure sectors

4.1. Assess the Vulnerability of Critical Infrastructure Systems to Space Weather

To enable and increase situational awareness during an extreme space-weather event, Federal agencies will coordinate with academia and the private sector to accomplish the following action:

4.1.1. DHS, in collaboration with Sector Specific Agencies (SSAs), will assess the vulnerability of critical infrastructure to space-weather events (as described in Goal 1). The assessment will include interdependencies and failure modes among sectors that can lead to cascading failures and will identify gaps where scientific or engineering research is required to understand or mitigate risks to critical infrastructure. The assessments will use the Phase 1 benchmarks as an initial input and will be reevaluated upon the completion of Phase 2 benchmarks.

Deliverable: Complete assessment reports

Timeline: The initial assessments will be completed within 18 months of the development of Phase 1 benchmarks. Reevaluations based on the Phase 2 benchmarks will be completed within 1 year of the development of Phase 2 benchmarks

4.2. Develop a Real-Time Infrastructure Assessment and Reporting Capability

The following actions will enable and increase capacity for real-time monitoring of the electric power system during space-weather events:

4.2.1. DOE, in coordination with DHS, DOC, and stakeholders in the energy sector, will develop plans to provide monitoring and data collection systems. The plans will inform a systemwide, real-time view of geomagnetically induced currents (GICs) at the regional level and, to the extent possible, display the status of power generation, transmission, and distribution systems during geomagnetic storms.

Deliverable: Complete plan for national GIC and grid monitoring system and delineate responsibilities for deployment

Timeline: Within 1 year of the publication of this Action Plan

4.2.2. DOE, in coordination with regulatory agencies and the electric power industry, will define data requirements that facilitate a centralized reporting system to collect real-time information on the status of the electric power transmission and distribution system during geomagnetic storms.

Deliverable: Define data requirements

Timeline: Within 1 year of the publication of this Action Plan

The following actions will define requirements for real-time assessment and reporting of radiation impacts to aviation safety:

4.2.3. DOC, in coordination with NASA, DOD, and DOT, will work with the commercial aviation industry, space operations and services, and international groups to define the requirements for real-time monitoring of the charged particle radiation environment to protect the health and safety of crew and passengers during space-weather events.

Deliverable: Complete document on radiation monitoring requirements

Timeline: Within 1 year of the publication of this Action Plan

4.2.4 DOT, in coordination with DOC, the Department of State (DOS), NASA; and in collaboration with commercial aviation, space, and international stakeholders; will define the scope and requirements for a real-time reporting system that conveys situational awareness of the radiation environment to orbital, suborbital, and commercial aviation users during space-weather events.

Deliverable: Develop and implement the mechanism for communicating real-time radiation awareness to aviation operators

Timeline: Within 2 years of the publication of this Action Plan

4.2.5. DOC and DOT, in coordination with NASA, academia, the private sector, and international partners, will develop or improve models for the real-time assessment of radiation levels at commercial flight altitudes.

Deliverable: Develop commercial aviation radiation-environment models ready for operational transition

Timeline: Within 2 years of the publication of this Action Plan

The following actions define requirements for real-time assessment and reporting of impacts to radio and satellite communications and space-based PNT systems:

4.2.6. DOC, in coordination with NSF and DOI, and commercial communication and PNT system stakeholders, will define requirements for real-time monitoring systems to assess atmospheric conditions that could affect these systems during ionospheric disturbances and geomagnetic storms.

Deliverable: Define requirements for a national operational network of real-time ionospheric monitoring stations

Timeline: Within 1 year of the publication of this Action Plan

4.2.7. DOC, DOD, and DHS, in coordination with government and commercial communications and PNT system users, will define the scope and observational requirements for a system that provides near-real-time situational awareness of the space environment for communication and PNT systems.

Deliverable: Complete report with scope and observational requirements

Timeline: Within 1 year of the publication of this Action Plan

4.2.8. DOC and DOD will create and support a satellite-anomaly database to enable secure collection and analysis of satellite-anomaly data related to space weather.

Deliverable: Complete development of a satellite-anomaly database in a secure format at DOC

Timeline: Within 1 year of the publication of this Action Plan

4.3. Develop or Refine Operational Models that Forecast the Effects of Space Weather on Critical Infrastructure

Accurate forecasts of the effects of a space-weather event on critical infrastructure require numerical models that can inform forecasters, decision-makers, and emergency managers prior to and during an event. The following actions are essential to achieving this objective:

4.3.1. DHS, in coordination with SSAs, will work with forecasting centers, emergency managers, governments, and academic and commercial stakeholders to define and develop comprehensive requirements for operational models to forecast the effects of space weather on critical infrastructures.

  Deliverable: Define sector-specific requirements for developing operational models for the effects of space weather on critical infrastructures

  Timeline: Within 2 years of the publication of this Action Plan

4.3.2. DHS and DOC, in coordination with SSAs and stakeholders, will identify gaps in current modeling capabilities and work with the research community to develop new and improved impact models and decision support tools. This will include a survey of current infrastructure impact models to determine if these models adequately account for the effects of space-weather events.

  Deliverable: Complete survey of existing impact models

  Timeline: Within 1 year of the completion of Action 4.3.1

4.3.3. Taking into account the results of the survey of existing models called for in Action 4.3.2, DHS and DOC, in coordination with SSAs, and in collaboration with the research community and stakeholders, will test and validate the existing suite of infrastructure impact models (developed by government, private-sector, and academic stakeholders) that enable forecasting of the full range of interconnected effects during an extreme space-weather event. New models will be developed if gaps are identified.

  Deliverable: Complete testing and validation of existing models and provide a plan to address any identified gaps

  Timeline: Within 1 year of the completion of Action 4.3.2

4.3.4. DHS, in coordination with DOC and SSAs, will incorporate infrastructure impact models into existing and future exercises to develop realistic space-weather scenarios for response and recovery, including societal impacts.

  Deliverable: Complete extreme space-weather simulations on a national scale and complete an analysis of results for full system performance

(including event forecasting, communications, impacts forecasting, mitigation, and response)
Timeline: Within 1 year of the completion of Phase 1 benchmarks

4.3.5. DHS and DOC, in coordination with government, private sector, and academic stakeholders, will recommend a policy for standardizing communication and data formatting from infrastructure monitoring systems and model outputs.
Deliverable: Complete recommended policy on data access
Timeline: Within 1 year of the publication of this Action Plan

4.3.6. DHS and DOC will develop data stewardship, archiving, and access-provision capabilities for space-weather infrastructure impact data and model output.
Deliverable: Develop policies on stewardship, archiving, and access of data
Timeline: Within 1 year of the completion of 4.3.5

4.4. Improve Operational Impact Forecasting and Communications

An operational capability that can forecast the effects of space weather is required to enable timely warnings to policy makers, system operators, and emergency managers. The following actions will be taken to establish this capability:

4.4.1. DOC and DHS, in coordination with other relevant agencies and stakeholders, will conduct a survey of commercial systems operators, government operators, and emergency managers to identify and assess the requirements for developing functional forecasting capabilities and alert products, including specifications for lead time, accuracy, and uncertainty.
Deliverable: Complete documentation of forecast content and lead-time requirements for relevant critical infrastructure sectors
Timeline: Within 1 year of the publication of this Action Plan

4.4.2. DHS and DOC, in coordination with NASA, NSF, private sector, academia, and other stakeholders, will develop a national capability for operational forecasting of space-weather impacts. The process will seek the development of new or improved forecasting models and the development of relevant tools and products that ensure the operational execution and dissemination of forecasts.

Measure of performance: Complete the development of new or improved operational forecasting models for at least two critical infrastructure sectors: energy and communications

Timeline: Within 3 years of the publication of this Action Plan

4.5. Conduct Research on the Effects of Space Weather on Industries, Operational Environments, and Infrastructure Sectors

The fundamental physical characteristics of space-weather effects on critical infrastructure, such as the electric power system, are not fully understood. The following actions are intended to address gaps in scientific, engineering, social, and economic understanding of these characteristics:

4.5.1. DHS, in coordination with SSAs, will support scientific and engineering research by governments, academia, and private-sector stakeholders to increase understanding of space-weather effects on critical infrastructure systems and to develop measurement systems and tools that enhance the forecasting and mitigation of effects.

Deliverable: Complete review of the extent to which grant programs at various agencies support research on the effects of space weather on critical infrastructure, and identify opportunities to introduce new programs or enhance existing processes

Timeline: Within 1 year of the publication of this Action Plan

4.5.2. DOC, in coordination with DHS, will support research into the social and economic impacts of space-weather effects, including costs of addressing irregularities in the electric power distribution system, addressing airline radiation, and lost productivity due to Global Navigation Satellite System signal impacts. Agencies will develop quantitative estimates of the potential costs of a space-weather event at the most severe level of estimated impact.

Deliverable: Initiate studies on the economic impact of space-weather events

Timeline: Within 2 years of the publication of this Action Plan

GOAL 5: IMPROVE SPACE-WEATHER SERVICES THROUGH ADVANCING UNDERSTANDING AND FORECASTING

Introduction

Space-weather information products are distributed to users worldwide and are vital for protecting life and property and for promoting economic productivity. Accurate, understandable, and timely space-weather information enables actions that reduce the vulnerability of interdependent national critical infrastructure. Building hazard-resilient communities requires an underlying network of interconnected resilient technologies and critical infrastructures. Understanding, predicting, and managing the effects of space weather—extreme events in particular—presents many challenges and requires continued investment in observations, modeling, and forecasting.

Observations are the backbone of forecast and warning capabilities. To achieve a robust operational program for space-weather observations, the United States must: (1) establish and sustain a foundational set of observations; (2) when feasible and cost effective, use data from multiple sources, including international, Federal, State, and local governments, as well as from the academic and industry sectors; (3) ensure the continuity of critical data sources; (4) continue to support sensors for solar and space physics research; (5) ensure data-assimilation techniques are in place; and (6) maintain archives for ground- and space-based data, which are essential for model development and benchmarking.

Both applied and basic research are necessary to improve space-weather services. Applied research must have near-term goals and focus on needed services and technologies. Basic research activities can advance the understanding of the dynamic processes of the sun and the sun-Earth connection.

Enhanced understanding of these processes will lead to improved space-weather forecasts, warnings, and mitigation efforts. Federal and non-Federal partners must ensure that research is effectively transitioned to operational forecasting centers (e.g., the National Oceanic and Atmospheric Administration's Space Weather Prediction Center), meeting the needs of these centers and other users.

Timely and accurate space-weather information products will ensure that emergency managers, first responders, government officials, businesses, and the public will be empowered to make fast, smart decisions in response to space-weather events.

These efforts must be closely coordinated and mutually supportive to efficiently and effectively meet the growing need for the delivery of space-weather information and services through collaborations among the Federal, private-sector, academic, and international communities. These collaborations can enhance the Nation's research, effectively transition research to operations, and provide the services needed to protect critical infrastructure.

The objectives for Goal 5 are:

- Improve understanding of user needs for space-weather forecasting to establish lead-time and accuracy goals
- Ensure that space-weather products are intelligible and actionable to inform decision-making
- Establish and sustain a baseline observational capability for space-weather operations
- Improve forecasting lead-time and accuracy
- Enhance fundamental understanding of space weather and its drivers to develop and continually improve predictive models
- Improve effectiveness and timeliness of the process that transitions research to operations

5.1. Improve Understanding of User Needs for Space-Weather Forecasting to Establish Lead-Time and Accuracy Goals

5.1.1. DOC will conduct a comprehensive survey of space-weather data and product requirements needed by user communities to help improve services.

Deliverable: Complete survey and associated analysis of user requirements

Timeline: Within 1 year of the publication of this Action Plan

5.2. Ensure Space-Weather Products Are Intelligible and Actionable to Inform Decision-Making

5.2.1. DOC and DOD, in coordination with DHS, will assess best practices across the Federal Government to identify and document the most effective means to produce and deliver space-weather alerts, warnings, and notifications.

Deliverable: Complete report on best practices with recommendations for implementation

Timeline: Within 6 months of the publication of this Action Plan

5.2.2. DHS, in coordination with DOC, will develop space-weather event-specific protocols that define the chain of command, control, and communication of space-weather-impact information during an extreme space-weather event.

Deliverable: Determine space-weather event-specific protocols

Timeline: Within 1 year of the publication of this Action Plan

5.3. Establish and Sustain a Baseline Observational Capability for Space-Weather Operations

To ensure that an extreme space-weather event is detected before it affects Earth, and to enable future improvements while maintaining current levels of products and services, the United States must establish and sustain a set of baseline space- and ground-based observations. These platforms must meet reliability standards to ensure the observing systems reliably deliver the data and data-derived products. The associated data reception, relay, processing, assimilation, and archiving infrastructure required to utilize space-weather observations must also be included in the baseline.

The following two actions are priorities to sustain current operational observing capabilities:

5.3.1. DOC, NASA, and NSF will develop a strategy for: (1) the continuous operation of the Solar and Heliospheric Observatory/Large Angle and Spectrometric Coronagraph (SOHO/LASCO) for as long as the satellite continues to deliver quality observations; and (2) prioritizing the reception of LASCO data in anticipation of extreme space-weather events.

Deliverable: Complete strategy to sustain SOHO/LASCO operations

Timeline: Within 1 year of the publication of this Action Plan

5.3.2. DOC, in coordination with NASA and DOD, will develop options to deploy an operational satellite mission to a position at least 1 million miles upstream on the sun-Earth line (e.g., the L1 Lagrangian point). The primary instrument on this mission will be a solar coronagraph to replace the SOHO/LASCO coronagraph capability. This mission will also provide solar wind measurements and other measurements essential to space-weather forecasting.

Deliverable: Complete an analysis of alternatives to achieve operational status in time to ensure continuity of coronagraph and solar wind data from the L1 Lagrangian point. This analysis will also consider commercial solutions and international partnerships.

Timeline: By the end of 2017

The following actions will be taken to establish and sustain baseline operational observing capabilities:

5.3.3. DOC will sustain or enhance solar imaging and measurements of solar X-ray irradiance, energetic particles, and *in situ* magnetic field vectors from geostationary orbit.

Deliverable: Achieve sustained measurement and data continuity
Timeline: Continuous

5.3.4. DOC and DOD, in coordination with NSF, will sustain or enhance ground-based solar imaging, including solar magnetic field and H-alpha data for operational forecasting.

Deliverable: Achieve sustained measurement and data continuity for at least 10 years
Timeline: Continuous

5.3.5. DOD, in coordination with DOC, will sustain or enhance ground-based solar radio capabilities that provide continuous observations of solar radio emission to operational forecasting centers.

Deliverable: Achieve sustained measurements from ground-based solar radio capabilities for at least 10 years
Timeline: Continuous

5.3.6. DOI, in coordination with DOC, will sustain the existing ground-based geomagnetic monitoring network and enhance the network through the installation of new observatories that will deliver data to operational centers in real time.

Deliverable: Achieve sustained measurement and data continuity, complete installation of new observatories, develop a real-time data delivery system, and develop enhanced visualization and analysis tools for geomagnetic activity

Timeline: Continuous for sustained measurement; enhancements completed within 5 years of the publication of this Action Plan and sustained thereafter

5.3.7. DOC and DOD will enable and sustain the acquisition and delivery of satellite-based Global Navigation Satellite System radio occultation data with sufficient geographical coverage, data-rate, and latency to satisfy operational ionospheric-forecasting requirements. DOC will also ensure that such data are assimilated into operational models of Earth's thermosphere and ionosphere.

Deliverable: Achieve sustained acquisition, delivery, and assimilation of data for operational models

Timeline: Data acquired and assimilated by 2018

5.3.8. DOC, DOD, and NSF, in collaboration with academia, the private sector, and international partners, will develop options to sustain or enhance the worldwide ground-based neutron-monitoring network to include real-time reporting of ground-level events to operational space-weather-forecasting centers.

Deliverable: Complete plan to ensure a sufficient number of neutron detectors are deployed, worldwide, to adequately characterize the radiation environment and support a real-time alert and warning system

Timeline: Within 6 months of the publication of this Action Plan

The following actions will be taken to prioritize and plan the establishment of the baseline operational observing system:

5.3.9. DOC, in coordination with NASA, DOD, and NSF, will produce a plan for deployment of new operational space-weather-observing assets to provide the baseline measurements outlined above. The plan will prioritize and define the required fidelity, cadence, and latency of ground-based and space-based measurements.

Deliverable: Complete prioritized plan to provide baseline space-weather observations

Timeline: Within 2 years of the publication of this Action Plan

5.3.10. DOC, NASA, NSF, DOD, and DOI will develop a plan to sustain the availability of facilities for the calibration of space-weather-observing assets to ensure that measurements are accurate and comparable through traceability to international standards.

Deliverable: Complete plan to sustain the availability of facilities for the calibration of spaceweather-observing assets

Timeline: Within 1 year of the publication of this Action Plan

5.4. Improve Forecasting Lead-Time and Accuracy

Space-weather forecasters analyze near-real-time ground- and space-based observations to assess the current-state space environment. Forecasts are based on a mixture of observations and model calculations, with models relying on observations as inputs. Actions identified to improve both the accuracy and lead-time of space-weather forecasts include:

5.4.1. NASA and DOC will assess space-weather-observation platforms with deep-space orbital positions (including candidate propulsion technology), which allow for additional warning time of incoming space-weather events.

Deliverable: Complete assessment report

Timeline: Within 1 year of the publication of this Action Plan

5.4.2. NASA, DOC, DOD, and NSF will support the development of novel sensor technologies and instrumentation to improve forecasting lead-time and accuracy.

Deliverable: Complete assessment of technology needs

Timeline: Within 1 year of the publication of this Action Plan

5.4.3. NASA, DOC, DOD, and NSF will prioritize and identify needs for improved coverage, timeliness, data rate, and data quality for space-weather observations, and opportunities to address these needs through collaborations with academia, the private sector, and international community.

Deliverable: Develop a report with priorities and recommendations

Timeline: Within 1 year of the publication of this Action Plan and every year thereafter, as necessary

5.5. Enhance Fundamental Understanding of Space Weather and Its Drivers to Develop and Continually Improve Predictive Models

Accurate space-weather forecasts depend largely on understanding the complex interactions between the sun and the Earth. Limited understanding of these interactions hinders accurate forecasting of space-weather events. Additional effort is needed to improve the understanding of these sun-Earth interactions that produce space weather. Actions identified to advance these efforts include:

5.5.1. NSF and NASA, in collaboration with DOC and DOD, will lead an annual effort to prioritize and identify opportunities for research and development (R&D) to enhance the understanding of space weather and its sources. These activities will be coordinated with existing National-level and scientific studies. This effort will include modeling, developing, and testing models of the coupled sun-Earth system and quantifying the long- and short-term variability of space weather.

Deliverable: Document R&D priorities

Timeline: Within 1 year of the release of this Action Plan and every year thereafter, as necessary

5.5.2. NASA, NSF, and DOD will identify and support basic research opportunities that seek to advance understanding of solar processes and how the sun's activity connects to and drives changes on Earth and its near-space environment.

Deliverable: Announce and provide financial awards that enhance basic research in this area

Timeline: Within 1 year and sustained thereafter

5.5.3. NASA, DOC, and DOD will identify and support research opportunities that seek to address targeted operational space-weather needs.

Deliverable: Announce and provide awards that enhance research in focused areas

Timeline: Within 1 year and sustained thereafter

5.5.4. DOI will assess and pilot a geo-electric monitoring capability through the installation of sensors at existing observatories. Data from these observatories will enhance the validation of electric field models.

Deliverable: Complete and begin geo-electric monitoring pilot program
Timeline: Within 1 year of the publication of this Action Plan

5.5.5. DOI will identify and fill gaps in magnetotelluric (MT) surveys[19] of the United States, beginning with the northeastern United States and concentrating on geographic regions judged to have the highest induction hazards.

Deliverable: Complete improvements to localized estimates of geo-electric fields and in lithospheric conductivity models
Timeline: Within 1 year of the publication of this Action Plan, surveys of the northeastern United States will be completed; the remaining components of this action are a long-term effort

5.5.6. DOI will map geomagnetic and geo-electric hazards using observatory and MT data.

Deliverable: Complete map of geomagnetic and geo-electric hazards for targeted geographic regions
Timeline: Within 1 year of the publication of this Action Plan, complete real-time geomagnetic mapping project for DOC, perform scenario storm analysis of induction hazards at sites where MT surveys have been made, develop a report on the feasibility of providing DOC with a real-time service for geo-electric field maps. Within 2 years, using conductivity models derived from MT surveys, perform scenario storm analysis of induction hazards across specific, geographically continuous regions of the United States

5.6. Improve Effectiveness and Timeliness of the Process that Transitions Research to Operations

The ability to effectively transition research to sustained operations is a critical element for improving space-weather products and services. The following actions will facilitate the transition of needed space-weather information and prediction capabilities to the Nation's space-weather service providers:

5.6.1. NASA and NSF, in collaboration with DOC and DOD, will develop a formal process to enhance coordination between research modeling centers and forecasting centers. This process will seek to identify roles and responsibilities in testing, verification, and validation for transitioning space-weather research models to space-weather-forecasting centers and for sustaining and improving models that transition into operations.

Deliverable: Signed memorandum of understanding between modeling and forecasting centers

Timeline: Within 6 months of the publication of this Action Plan

5.6.2. DOC and DOD, in collaboration with NASA and NSF, will develop a plan (which may include a center) that will ensure the improvement, testing, and maintenance of operational forecasting models. This action will leverage existing capabilities in academia and the private sector and enable feedback from operations to research to improve operational space-weather forecasting.

Deliverable: Complete plan for improving, testing, and maintaining operational forecasting models and enabling operations-to-research feedback

Timeline: Within 6 months of the publication of this Action Plan

GOAL 6: INCREASE INTERNATIONAL COOPERATION

Introduction

Space weather is a worldwide threat and a concern shared by many nations. The technologies that have revolutionized the global economy and transformed lives are the same technologies that are vulnerable to space weather. This phenomenon is not constrained by national boundaries and has potential to affect many parts of the globe simultaneously. No one nation can meet this global challenge alone.

The United States must partner with other nations in developing and strengthening standards and protocols for the protection of key infrastructure. The actions in this chapter define an approach that unifies U.S. engagement with the international community. The Action Plan identifies the bilateral and multilateral cooperation necessary to promote safety, security, and economic stability before and after space-weather events.

This Action Plan recognizes that many important international initiatives are already underway. Goal 6 actions seek to support these efforts and encourage increased cooperation. The United States must consider complementary approaches to include internationally-negotiated top-down initiatives to further understanding and approaches to space-weather events.

The Nation must manage engagements with international partners under the authority of existing Presidential Directives and international agreements.

National efforts will, where possible, advance regulatory coherence and enhance efforts of international organizations.

Common solutions to regional challenges associated with space weather and exchange of best practices between the United States and international partners will strengthen global capacity to respond to extreme space-weather events. These actions pursue an international collaboration that will empower the United States and its international collaborators to be prepared to withstand the effects from space weather.

The objectives for Goal 6 are:

- Build international support and policies for acknowledging space weather as a global challenge
- Increase engagement with the international community on observation infrastructure, data sharing, numerical modeling, and scientific research
- Strengthen international coordination and cooperation on space-weather products and services
- Promote a collaborative international approach to preparedness for extreme space-weather events

6.1. Build International Support at the Policy Level for Acknowledging Space Weather as a Global Challenge

A prerequisite to enhanced international cooperation is high-level support across partner nations to raise awareness of space weather as a global challenge. The following actions will build international support at the policy level:

6.1.1. DOS, in coordination with other agencies, will ensure that policy makers and leaders of partner nations are informed of the need for a comprehensive and coordinated approach to preparing for an extreme space-weather event.

Deliverable: Organize and host a high-level international meeting on economic and societal effects of an extreme space-weather event

Timeline: Within 18 months of the publication of this Action Plan

6.1.2. DOS will coordinate sustained U.S. participation in relevant international space-weather initiatives. This effort will include participation in

United Nations activities and incorporation of space-weather-related elements into work plans, programs, and projects by supporting the following activities:

- Develop a 4-year plan for United Nations (UN) World Meteorological Organization (WMO) space-weather activities
- Continue including space weather as a regular agenda item of the Scientific and Technical Subcommittee of the UN Committee on the Peaceful Uses of Outer Space (COPUOS)
- Provide global space-weather information and services for international aviation with the UN International Civil Aviation Organization (ICAO)
- Provide guidance on ionospheric disturbances monitoring and forecasting with the International Telecommunications Union (ITU)

Deliverable: Complete report on progress and international engagement
Timeline: Within 1 year of the publication of this Action Plan and every year thereafter

6.2. Increase Engagement with the International Community on Observation Infrastructure, Data Sharing, Numerical Modeling, and Scientific Research

Increased access to government, civilian, and commercial space-weather observational infrastructure and data across the globe is of mutual benefit to the United States and its partners. Consistent with the U.S. Open Data Action Plan,[20] Federal agencies will facilitate full and open access to data to advance international cooperation in the characterization, prediction, and mitigation of space-weather effects. These same agencies should encourage international science and service partners to adopt policies that promote full and open access to data, consistent with the G8 Open Data Charter,[21] the international Group on Earth Observation (GEO) Data Sharing Principles,[22] and WMO Resolution 40 principles,[23] with an emphasis on real-time data access. The following actions will increase engagement with the international community on observation infrastructure, data sharing, numerical modeling, and scientific research:

6.2.1. DOI will lead development of a plan for expansion of the real-time ground-based magnetometer network to improve global geophysical monitoring.

   Deliverable: Complete strategy for expanding the magnetometer network

   Timeline: Within 1 year of the publication of this Action Plan

6.2.2. DOC and DOI, in coordination with NASA and NSF, will explore opportunities to leverage international partnerships to sustain baseline operational space-weather-observing capabilities.

   Deliverable: Complete report on international partnerships

   Timeline: Within 1 year of the publication of this Action Plan and every year thereafter

6.2.3. DOC and NASA will collaborate with academia, the private sector, and the international community to explore the potential benefits and costs of space-weather missions in orbits complementary to the sustained missions at the L1 Lagrangian point, which may include missions at the L5 Lagrangian point. Such missions may improve monitoring of CME properties and trajectories relative to Earth.

   Deliverable: Complete analysis of space-weather missions in orbits complementary to sustained missions.

   Timeline: Within 1 year of the publication of this Action Plan

6.2.4. DOC, in collaboration with DOS, will sustain and enhance international partnerships for the acquisition of data from solar-imaging and solar-wind deep-space missions, building on the ongoing operational Real-Time Solar Wind (RTSW) network. [24]

   Deliverable: Complete report on the status of operational deep-space mission data acquisition and needs for additional antenna resources

   Timeline: Within 1 year of the release of this Action Plan

6.2.5. DOC, in coordination with DOI, will maintain U.S. input to the WMO Observing System Capability Analysis and Review (OSCAR) database and encourage contributions of international partners to ensure comprehensive knowledge of international space-weather observational systems and their data products currently in use and planned for operational forecasting. This action will include information on ground- and space-based systems.

   Deliverable: Complete documentation of international observational systems

Timeline: Within 1 year of the publication of this Action Plan, with annual updates

6.2.6. DOC and DOI, in coordination with NSF and NASA, will promote the improved exchange of data and information using the WMO Information System and other means, and organize international data comparison activities to promote the availability, intercalibration, and interoperability of space- and ground-based data.

Deliverable: Complete report on the use of the WMO Information System and other data-sharing means and on international data comparison activities

Timeline: Within 2 years of the publication of this Action Plan

6.2.7. DOC and DOI, in coordination with NSF and NASA, will provide input to the WMO operational space-weather-observing requirements and Statement of Guidance and will report to relevant international organizations, including the COPUOS, the Coordination Group for Meteorological Satellites (CGMS), and the International Real-time Magnetic Observatory Network (INTERMAGNET),[25] on priorities for coordinated action.

Deliverable: Submit report to each of the noted international organizations at their respective primary annual meetings

Timeline: Within 1 year of the publication of this Action Plan and every year thereafter

6.2.8. NASA will promote and support the continuation of space weather as a regular topic in the international efforts of the International Council for Science's Committee on Space Research (COSPAR) and within the International Living with a Star (ILWS) program.

Deliverable: Complete progress report

Timeline: Within 1 year of the publication of this Action Plan and every year thereafter

6.3. Strengthen International Coordination and Cooperation on Space-Weather Products and Services

Providing high-quality space-weather products and services worldwide requires international consensus and cooperation. To achieve this, the following actions are necessary: international agreement on common terminology, measurements, and scales of magnitude; promotion, coordination, and dissemination of space-weather observations, model outputs, and forecasts; and establishment of coordination procedures across space-weather operations centers during extreme events. The following actions will strengthen international cooperation on space-weather products and services:

6.3.1. DOC will lead U.S. efforts to engage international partners to ensure that communicated products and services are globally consistent during extreme events.

Deliverable: Complete report on actions taken to sustain, improve, and develop international coordination mechanisms

Timeline: Within 1 year of the publication of this Action Plan and every year thereafter

6.3.2. DOT, in coordination with DOC and DOD, will lead U.S. efforts to develop international standards for the provision of space-weather information for international air navigation.

Deliverable: Develop proposal for ICAO
Timeline: Within 1 year of the publication of this Action Plan

6.3.3. DOC and NASA will continue efforts within CGMS to promote an ongoing agenda item on space-weather activities.[26]
Deliverable: Complete progress report
Timeline: Within 1 year of the publication of this Action Plan and every year thereafter

6.3.4. DOC will sustain engagement with the International Space Environment Service (ISES) and foster participation of additional nations in the network of space-weather service providers.
Deliverable: Complete progress report
Timeline: Within 1 year of the publication of this Action Plan and every year thereafter

6.4. Promote a Collaborative International Approach to Preparedness for Extreme Space-Weather Events

The world's interconnected and interdependent systems are vulnerable to extreme space-weather events, which could possibly lead to a cascade of effects across borders and sectors. To mitigate these risks, the United States will work with the international community to facilitate the exchange of information and best practices to strengthen global preparedness capacity for extreme space-weather events. The United States will also foster development of global mutual-aid arrangements to facilitate response and recovery efforts, and will coordinate international partnership activities to support space-weather preparedness and response exercises. The following actions will promote a collaborative international approach to preparedness for extreme space-weather events:

6.4.1. DOS, DOC, and DHS, in coordination with and other Federal agencies, will provide outreach and education to assist nations in understanding space-weather effects and integrating space weather into national hazard and risk registries.

Deliverable: Complete progress report

Timeline: Within 1 year of the publication of this Action Plan and every year thereafter

6.4.2. DOS, DHS, NSF, DOE, and DOC will work with relevant international organizations[27] and key partners on assessing global economic impact of an extreme space-weather event.

Deliverable: Develop proposal for international assessment

Timeline: Within 1 year of the publication of this Action Plan

6.4.3. The United States Postal Service (USPS), DOT, and DHS will participate in the Civil Emergency Planning Committee (CEPC) of the North Atlantic Treaty Organization (NATO) to advise NATO planners on possible implications of space weather for NATO operations; to promote consistency in communications and operations among NATO members and partner nations; and to assist in and, as appropriate, lead development of training and exercise events.

Deliverable: Complete progress report

Timeline: Within 1 year of the publication of this Action Plan and every year thereafter

6.4.4. DOS, in coordination with DHS, DOD, and DOC, will develop space-weather event-specific protocols that define the communication of U.S. space-weather-impact information to other nations and international organizations during an extreme space-weather event.

Deliverable: Complete development of communications protocols
Timeline: Within 2 years of the publication of this Action Plan

6.4.5. DOS, in coordination with DHS, DOC, and DOT, will inform U.S. Embassies and Missions worldwide of the effects from an extreme space-weather event.
    Deliverable: Complete outreach strategy
    Timeline: Within 6 months of the publication of this Action Plan

6.4.6. DOS, in coordination with relevant agencies, and consistent with Office of Management and Budget (OMB) Circular A-119[28] and Executive Office of the President (EOP) Memo M-12- 08,[29] will support the development and use of international standards for improved resilience of equipment to extreme space-weather events by participating in the development of relevant open, consensus-based international standards.
    Deliverable: Complete progress report
    Timeline: Within 1 year of the publication of this Action Plan and every year thereafter

6.4.7. DOS, in coordination with DHS, DOC, and other Federal agencies, will address extreme space-weather events in accordance with supply-chain issues and as part of the U.S. government's overall and ongoing efforts to implement the 2012 National Strategy for Global Supply Chain Security.
    Deliverable: Complete progress report
    Timeline: Within 1 year of the publication of this Action Plan and every year thereafter

CONCLUSION

The activities outlined in this Action Plan represent a merging of national and homeland security concerns with scientific interests. This effort is only the first step. The Federal Government alone cannot effectively prepare the Nation for space weather; significant effort must go into engaging the broader community. Space weather poses a significant and complex risk to critical technology and infrastructure, and has the potential to cause substantial economic harm. This Action Plan provides a road map for a collaborative and Federally-coordinated approach to developing effective policies, practices, and procedures for decreasing the Nation's vulnerabilities. By specifying tasks that will lead to improvements in forecasting, research, preparedness, planning, and domestic and international engagement, and identifying the responsible agencies, this Action Plan will help ensure that the Federal Government and its domestic and international partners are able to withstand and quickly recover from effects of extreme space-weather events.

REFERENCES

Coordination Group for Meteorological Satellites (CGMS). "CGMS High Level Priority Plan (HLPP), 2015– 2019.

Blanchard, B. Wayne. "Guide to Emergency Management and Related Terms, Definitions, Concepts, Acronyms, Organizations, Programs, Guidance, Executive Orders & Legislation: A Tutorial on Emergency Management, Broadly Defined, Past and Present." October 22, 2008.

Department of Homeland Security (DHS). "National Infrastructure Protection Plan." Last updated June 16, 2015.

___. *The Strategic National Risk Assessment in Support of PPD 8: A Comprehensive Risk-Based Approach toward a Secure and Resilient Nation.* December 2011.

___.."Sector-Specific Agencies." March 2, 2015.

Executive Office of the President (EOP). "Principles for Federal Engagement in Standards Activities to Address National Priorities." EOP Memo M-12-08. January 17, 2012.

Federal Emergency Management Agency (FEMA). "Emergency Support Function 15: Standard Operating Procedures." Last updated August 21, 2014.

___.."National Planning Frameworks." Last updated March 19, 2015. www.fema.gov/national-planning-frameworks.

___.. "National Preparedness Goal." Last updated March 19, 2015. www.fema.gov/national-preparedness-goal.

___.. "Recovery Federal Interagency Operation Plan (FIOP)." Last updated July 30, 2014.

___.. "Whole Community." Last updated April 16, 2015.

___.. "A Whole Community Approach to Emergency Management: Principles, Themes, and Pathways for Action," FDOC 104-008-1. December 2011.

Federal Energy Regulatory Commission (FERC). "Reliability Standards for Geomagnetic Disturbances." Order No. 779, 143 FERC ¶ 61,147. May 23, 2013.

___.. "Reliability Standard for Geomagnetic Disturbance Operations." Order No. 797, 147 FERC 61,209. June 25, 2014.

___.. "Reliability Standard for Transmission System Planned Performance for Geomagnetic Disturbance Events." 151 FERC 61,134. May 14, 2015.

G8. "Open Data Charter." Policy Paper. June 18, 2013.

Group on Earth Observations (GEO). "GEO Data Sharing Principles Implementation."

International Real-time Magnetic Observatory network (INTERMAGNET). "Welcome to INTERMAGNET."

National Research Council. *Severe Space Weather Events—Understanding Societal and Economic Impacts, A Workshop Report.* Committee on the Societal and Economic Impacts of Severe Space Weather Events: A Workshop. 2008.

___. *Solar and Space Physics: A Science for a Technological Society.* 2013–2022 Decadal Survey in Solar and Space Physics. 2013.

National Science and Technology Council, *National Strategy for Civil Earth Observations.* April 2013.

National Space Weather Program Council. *Report on Space Weather Observing Systems: Current Capabilities and Requirements for the Next Decade.* Office of the Federal Coordinator for Meteorological Services and Supporting Research. 2013.

Office of the Federal Coordinator for Meteorological Services and Supporting Research. *Report of the Assessment Committee for the National Space Weather Program.* FCM-R24-2006, June 2006,

Office of Management and Budget. "Federal Participation in the Development and Use of Voluntary Consensus Standards and in Conformity Assessment Activities." OMB Circular A-119. Washington, DC: OMB, February 10, 1998.

Organisation for Economic Co-operation and Development (OECD). *OECD Futures Project on 'Future Global Shocks': Geomagnetic Storms.* Paris, France, 11 January 2011.

___.. *OECD Reviews of Risk Management Policies: Future Global Shocks–Improving Risk Governance.* Preliminary Version. Paris, France, n.d.

___. "International Futures Program."

Presidential Policy Directive/PPD-8. "National Preparedness." March 30, 2011.

Presidential Policy Directive/PPD-21. "Critical Infrastructure Security and Resilience." February 12, 2013.

Public Law 111-267. *The National Aeronautics and Space Administration Authorization Act of 2010*. October 11, 2010.

Schrijver, Carolus J., Kirsti Kauristie, Alan D. Aylward, Clezio M. Denardini, Sarah E. Gibson, Alexi Glover, Nat Gopalswamy, Manuel Grande, Mike Hapgood, Daniel Heynderickx, Norbert Jakowski, Vladimir V. Kalegaev,Giovanni Lapenta, Jon A. Linker, Siqing Liu, Cristina H. Mandrini, Ian R. Mann,Tsutomu Nagatsuma, Dibyendu Nandi, Takahiro Obara, T. Paul O'Brien,Terrance Onsager, Hermann J. Opgenoorth, Michael Terkildsen, Cesar E. Valladares, and Nicole Vilmer. "Understanding Space Weather to Shield Society: A Global Road Map for 2015–2025 Commissioned by COSPAR and ILWS" *Advances in Space Research* 55 (2015): 2745–2807.

World Meteorological Organization (WMO). "Resolution 40 (Cg-XII)."

U.S. Department of Energy, Office of Electricity Delivery and Energy Reliability Infrastructure Security and Energy Restoration. *Insurance as a Risk Management Instrument for Energy Infrastructure Security and Resilience*. 2013.

White House. *U.S. Open Data Action Plan*. May 9, 2014.

___.. *The National Space Policy of the United States of America*. June 28, 2010.

___.. *National Space Policy of the United States of America*. June 28, 2010.

ABBREVIATIONS

CEPC	Civil Emergency Planning Committee
CGMS	Coordination Group for Meteorological Satellites
CME	coronal mass ejection
COPUOS	Committee on the Peaceful Uses of Outer Space
COSPAR	Committee on Space Research
DHS	Department of Homeland Security
DOC	Department of Commerce
DOD	Department of Defense
DOE	Department of Energy
DOI	Department of the Interior
DOS	Department of State
DOT	Department of Transportation
E-field	induced geo-electric field at Earth's surface
EOP	Executive Office of the President

EPFAT	Emergency Power Facility Assessment Tool
ESA	European Space Agency
ESFLG	Emergency Support Function Leadership Group
EUV	extreme ultraviolet
FCC	Federal Communications Commission
FIOP	Federal Interagency Operations Plan
GEO	Group on Earth Observation
GIC	geomagnetically induced current
GPS	Global Positioning System
HHS	Department of Health and Human Services
ICAO	International Civil Aviation Organization
ILWS	International Living with a Star
INTERMAGNET	International Real-time Magnetic Observatory Network
ISES	International Space Environment Service
ITU	International Telecommunications Union
LASCO	Large Angle Spectrometric Coronagraph
LEO	low-Earth orbit
MT	magnetotelluric
NASA	National Aeronautics and Space Administration
NATO	North Atlantic Treaty Organization
NIST	National Institute of Standards and Technology
NSF	National Science Foundation
NSTC	National Science and Technology Council
OECD	Organisation for Economic Co-operation and Development
OMB	Office of Management and Budget
OSCAR	Observing System Capability Analysis and Review
PNT	position, navigation, and timing
POIA	Power Outage Incident Annex
PPD	Presidential Policy Directive
R&D	research and development
RTSW	Real-Time Solar Wind
RSFLG	Recovery Support Function Leadership Group
SCC	Sector Coordinating Council
SLTT	State, local, tribal, territorial
SNRA	Strategic National Risk Assessment
SOHO	Solar and Heliospheric Observatory
SRB	solar radio burst

SSA	Sector-Specific Agency
UN	United Nations
USPS	United States Postal Service
UV	ultraviolet
WMO	World Meteorological Organization

End Notes

[1] The effects of space weather on technology can extend beyond the near-Earth environment, but this issue is considered outside of the scope of this Action Plan.

[2] The ionosphere is a portion of the upper atmosphere 50 to 600 miles above the Earth's surface that can be affected by solar flares, coronal mass ejections, and the solar wind.

[3] Radio signals between 3 and 30 MHz are defined as high frequency.

[4] Atmospheric drag refers to the air resistance in the regions where satellites orbit.

[5] Refer to PPD-8 and the Federal Emergency Management Agency (FEMA) website. "National Planning Frameworks," last updated March 19, 2015, www.fema.gov/national-planning-frameworkswww.fema.gov/national-planning-frameworks.

[6] Local governments include tribal, territorial, and insular area governments.

[7] Whole Community partners refer to the Nation's larger collective emergency management team and include not only DHS and its partners at the Federal level, but also State, local, tribal, and territorial (SLTT) partners, non-governmental organizations such as faith-based and nonprofit groups and private sector industry, and individuals, families and communities. See FEMA website, "Whole Community," last updated April 16, 2015, www.fema.gov/whole-community.

[8] The FIOPs, one for each preparedness mission area, describe how the Federal Government aligns resources and delivers core capabilities.

[9] See FEMA website, "National Preparedness Goal," last updated September 2015, www.fema.gov/national-preparednessgoal.

[10] See FEMA website, "Recovery Federal Interagency Operation Plan (FIOP)," last updated July 2014, www.fema.gov/medialibrary/assets/documents/97360.

[11] See FEMA website, "Emergency Support Function 15: Standard Operating Procedures," last updated August 21, 2014, www.fema.gov/media-library/assets/documents/34369.

[12] External Affairs ensures that sufficient Federal assets are deployed to the field during incidents requiring a coordinated Federal response to provide accurate, coordinated, timely, and accessible information to affected audiences (e.g., governments, media, the private sector, and the local populace, including the special-needs population).

[13] SSAs are defined by PPD-21 for each of 16 critical infrastructure sectors. See DHS website, "Sector-Specific Agencies," last updated March 2, 2015, www.dhs.gov/sector-specific-agencies.

[14] The Homeland Infrastructure Threat and Risk Analysis Center developed the National Risk Estimate product line in 2010 to provide authoritative, coordinated, risk-informed assessments of key national security issues in the Nation's infrastructure protection community.

[15] Disaster resilience refers to the capability to prevent, or protect infrastructure from, significant multi-hazard threats and incidents and to expeditiously recover and reconstitute critical

services with minimum damage to public safety and health, the economy, and national security.

[16] See DHS website, "National Infrastructure Protection Plan," last updated June 16, 2015, www.dhs.gov/nationalinfrastructure-protection-plan.

[17] An operational forecasting capability is one that is available at any time, is resistant to all failure modes, and has rapid computation and dissemination mechanisms.

[18] The focus of support for aviation has been limited to conventional aviation altitudes, but commercial space travel is expected to expand in the coming decades and must be considered. The safety requirements for conventional aviation and space travel have different magnitudes of exposure to extreme space-weather events.

[19] A method to measure the subsurface electrical conductivity of Earth's crust.

[20] See White House, *U.S. Open Data Action Plan,* May 9, 2014.

[21] See G8, "Open Data Charter," 18 June 2013.

[22] See GEO website, "GEO Data Sharing Principles Implementation," www.earthobservations. org/geoss_dsp.shtml.

[23] See World Meteorological Organization (WMO), "Resolution 40 (Cg-XII)," www.wmo.int/ pages/about/Resolution40_en.html.

[24] For a list of international and domestic partners, see the Space Weather Prediction Center website, www.swpc.noaa.gov/products/ace-ground-station-tracking-plots.

[25] The INTERMAGNET program exists to establish a global network of cooperating digital magnetic observatories. See INTERMAGNET website, www.intermagnet.org/.

[26] CGMS identified space weather coordination as the highest priority for the period 2014–18. See CGMS website, "Coordination Group for Meteorological Satellites (CGMS)," http://www.eumetsat.int/website/home/AboutUs/InternationalCooperation/CoordinationGro upforMeteorologicalSatellites CGMS/index.html.

[27] For example, see *OECD Futures Project on 'Future Global Shocks': Geomagnetic Storms*, 14 January 2011.

[28] OMB, "Federal Participation in the Development and Use of Voluntary Consensus Standards and in Conformity Assessment Activities," OMB Circular A-119, February 10, 1998.

[29] EOP, "Principles for Federal Engagement in Standards Activities to Address National Priorities," EOP Memo M-12-08, January 17, 2012.

In: Space Weather
Editor: Peter Burton

ISBN: 978-1-63484-440-6
© 2016 Nova Science Publishers, Inc.

Chapter 3

REPORT ON SPACE WEATHER OBSERVING SYSTEMS: CURRENT CAPABILITIES AND REQUIREMENTS FOR THE NEXT DECADE[*]

Office of the Federal Coordinator for Meteorological Services and Supporting Research

PREFACE

In April 2011, the Office of Science and Technology Policy (OSTP) in the Executive Office of the President asked the Office of the Federal Coordinator for Meteorological Services and Supporting Research (OFCM), under the auspices of the National Space Weather Program Council (NSWPC), to lead a study to assess (1) the current and planned space weather observing systems and (2) the capacity of those systems to meet operational space weather forecasting requirements over the next 10 years.

The request from OSTP followed passage of the NASA Authorization Act of 2010, which directed OSTP to arrange for such an assessment and report the results to appropriate Congressional committees. The NSWPC formed an interagency Joint Action Group (JAG) to execute the study, comprising 25 people from 15 Federal offices. In August 2011, the JAG briefed the NSWPC on the interim results of the study, with OSTP and the Office of Management

---

[*] This is an edited, reformatted and augmented version of a report issued by the Office of Science and Technology Policy, Executive Office of the President, April 2013.

and Budget (OMB) present as observers. This report, which formally documents the study results, was reviewed and approved by all interagency NSWPC members.

This report describes the study process, the study requirements and their relevance and importance, an assessment and accounting of current and planned space weather observing systems used or to be used for operations, an analysis of gaps between the observing systems' capabilities and their ability to meet documented requirements, and a summary of key findings. The report provides OSTP with a consolidated consensus view of the National Space Weather Program Federal agency partners with regard to key capabilities that need to be maintained, replaced, or upgraded to ensure space weather observing systems can meet the requirements of the Nation's critical space weather forecasting capabilities for the next 10 years. Of course, specific program activities are subject to future budgetary decisions.

The National Space Weather Program is a Federal interagency initiative with the mission of advancing the improvement of space weather services and supporting research in order to prepare the country for the technological, economic, security, and health impacts that may arise from extreme space weather events. The goal of the program is to achieve an active, synergistic, interagency system able to provide timely, accurate, and reliable space weather, observations, warnings, analyses, and forecasts.

I want to thank the JAG for its excellent service crafting this report. Special praise is due to the group's co-chairs, Dr. Bill Denig and Colonel John Egentowich, whose strong leadership ensured the success of this difficult undertaking.

Samuel P. Williamson
Federal Coordinator for Meteorological Services and Supporting Research
Chair, National Space Weather Program Council

EXECUTIVE SUMMARY

The 2010 National Aeronautics and Space Administration (NASA) Authorization Act, Section 809 (see Appendix 1) acknowledges:

- the threat to modern systems posed by space weather events;
- the potential for "significant societal, economic, national security, and health impacts" due to space weather disruptions of electrical power,

satellite operations, airline communications, and position, navigation, and timing systems; and

- the key role played by ground-based and space-based space weather observing systems in predicting space weather events.

In addition, the Act directed the Office of Science and Technology Policy (OSTP) to submit a report to the appropriate Congressional committees that details the following:

- "Current data sources, both space- and ground-based, that are necessary for space weather forecasting."
- "Space- and ground-based systems that will be required to gather data necessary for space weather forecasting for the next 10 years."

In response, OSTP requested the Office of the Federal Coordinator for Meteorological Services and Supporting Research (OFCM) on April 8, 2011, to lead the coordination of a new interagency assessment, under the auspices of the National Space Weather Program Council (NSWPC), to address the Act's requirements. The NSWPC established the Joint Action Group for Space Environmental Gap Analysis (JAG/SEGA) on April 28, 2011, to perform an assessment of existing and planned space weather observing systems and observing system requirements to support operational space weather forecasting over the next 10 years. On August 2, 2011, the JAG briefed interim results of the assessment to the NSWPC, with representatives of OSTP and the Office of Management and Budget (OMB) present as observers. This report is provided to satisfy OSTP's request as well as requirements of the Act.

The JAG/SEGA considered the following when defining the scope of the assessment documented in this report:

- **Requirements**: Proceed from currently documented observing requirements for operational space weather services.
  - Derived space weather observing requirements from those recently validated by Department of Defense (DoD), Department of Commerce (DOC) National Oceanic and Atmospheric Administration (NOAA) and NASA; hence, a revalidation of requirements was not needed.
  - Limited to observing requirements and systems necessary to drive operational forecasts and services. Pure research-only requirements were not considered.

- Requirements for observations needed to support space missions beyond Earth geosynchronous orbit (lunar, interplanetary, etc.) were also considered.
- **Observing Systems**: Use existing agency requirements, programs, initiatives, and plans for observing and forecasting systems.
  - Only existing or planned systems were considered. Potential new systems beyond those already planned were considered to be out of scope.
  - Operational systems and research platforms that can be leveraged for operational use were considered; research systems not suited for operational use were not considered.
  - International systems capable of supporting U.S. operational needs were considered.

The JAG/SEGA included 25 participants from 15 Federal organizations, representing the bulk of the U.S. Government space weather stakeholders. Representing the providers of the Nation's two primary operational space weather analysis and forecasting centers, leaders from the U.S. Air Force (USAF) and NOAA served as co-chairs for the JAG. Focusing on the specific goals set forth in the 2010 NASA Authorization Act, the JAG determined short-term and long-term space weather observing requirements needed to support operational space weather forecasting.

While the space weather observing requirements were specific to particular space weather environmental parameters, the JAG noted the importance of the requirements to the Nation's economy and security. As noted in the 2008 National Research Council (NRC) report, *Severe Space Weather Events*, "potential damage resulting from these critical dependencies [of critical infrastructure and systems to the space environment] can be minimized by having a robust capability to monitor, model, and predict what is happening in the space environment." Prominent potential impacts include:

- **Electric Power Grid:** Large scale blackouts and permanent damage to transformers, with lengthy restoration periods.
- **Global Satellite Communications:** Widespread service disruptions, which can impact financial, telemedicine, government, and Internet services, among many others.
- **GPS Positioning and Timing:** Degradations of military weapons accuracy, air traffic management, transportation, precision survey/construction/agriculture, energy exploration, ship

navigation/commerce, financial transactions, and cell phone/broadband.

- **Satellites & Spacecraft:** Loss of satellites and capabilities, loss of space situational awareness (including detection of hostile actions), increased probability of satellite-debris collisions, degraded communication/navigation, and increased risk to astronaut safety.

In assessing the existing and planned space weather observing systems needed to minimize the risk of these impacts and meet national requirements, the JAG considered ground-based and space-based solutions specifically designed for operations, research systems that are capable of being exploited for operations, and other domestic or international solutions that could be leveraged for operations. The JAG then used its compilation of the requirements, along with the existing and planned observing systems to be used to satisfy those requirements, and performed an analysis to determine key requirements shortfalls, or gaps ("gap analysis").

In conducting its analysis, the JAG noted that an observational requirement is a documented need for *measurements* of the space environment, which are contingent on the "domain" of the space environment in which the measurements are being made. For this assessment, observing requirements were categorized within the following six domains of the space environment: Sun/Solar, Heliosphere, Magnetosphere, Aurora, Ionosphere, and the Upper Atmosphere.

Within each of these six domains, several specific environmental parameters were identified and assessed against documented observing requirements. While the analysis of the ability of current, planned, and potential systems to meet specific observing requirements was critical to the assessment, the JAG took an additional step to ensure that the end results were tied to real-world applications. Specifically, the JAG mapped the observing parameters for each of the six domains to analysis and forecast products (nowcast, short-term forecast, and long-term forecast) for the five key space weather phenomena:

- **Geomagnetic Storms**: A worldwide disturbance of the Earth's geomagnetic field resulting from increases in the solar wind pressure and interplanetary magnetic field at the dayside magnetopause. The occurrence of substorms within a geomagnetic storm period can negatively impact satellite operations, power systems, radio propagation, and navigation systems.

- **Radio Blackouts**: Disturbances of the ionosphere caused by X-ray emissions from the Sun, which can negatively impact radio propagation and navigation systems.
- **Radiation Storms**: Elevated fluxes of charged particle radiation that can negatively impact satellite operations, radio propagation, navigation systems, and can increase biological risks to humans in spacecraft or high-flying aircraft.
- **Ionospheric Storms**: Disturbances in the ionosphere caused by large increases in the fluxes of solar particles and electromagnetic radiation, often associated with the occurrence of geomagnetic storms. There is a strong coupling between the ionosphere and the magnetosphere that often results in both regimes being disturbed concurrently. These disturbances can negatively impact radio communications as well as satellite navigation and communications systems.
- **Atmospheric Drag**: Collisions with diffuse air particles (altitudes typically < 2000 km) cause spacecraft to slow, leading them to gradually descend to lower altitudes where the drag continues to increase with increased atmospheric density. This phenomenon is affected by space weather since the density of the air particles responds to solar activity, such as magnetic storms. Solar emissions cause the upper atmosphere to heat and expand, which in turn increases drag at a given altitude. This effect increases dramatically with high solar activity. If the increased solar activity triggers increased magnetic activity at the Earth, intense currents, flowing through the upper atmosphere, also contribute to increased heating and expansion of the upper atmosphere. Accurate analysis of atmospheric drag effects can reduce the error associated with determination of satellite orbital intersection with other satellites and space debris, reducing the need for expenditure of fuel for orbital maneuvers and thereby extending the mission life of the spacecraft.

When consolidating the requirements and considering the ability of the current/planned systems to monitor the five key space weather phenomena included in the analysis, high-level impacts due to a few key systems become apparent. Table ES-1 (A) illustrates the degradation of operational capability should various key systems be lost due to launch/system failure, budget cuts, or other reasons; and (B) depicts the sustainment of current capabilities over time if key systems are maintained or replaced. It is particularly noteworthy that the addition of planned replacements or new systems maintains our current capabilities while providing some incremental improvement; none of these planned/replacement systems meet all requirements. Perhaps more importantly, this demonstrates the significant degradation in current capability should these planned/replacement systems not reach operational status. In other words, the Nation is at risk of losing critical capabilities that have significant economic and security impacts should these key space weather observing systems fail to be maintained and replaced.

Considering the rapidly growing dependency on space-based and space-enabled systems, which have permeated most facets of modern society, space weather observing and forecasting capabilities used to mitigate potential impacts will become even more critical in the future.

In performing the assessment of current and planned space weather observing systems and evaluating their ability to meet requirements, the JAG/SEGA arrived at the following key findings:

- A judicious mix of space-based and ground-based observing systems is currently used and needed to support operational space weather services.
  - The huge volume of the space environment means that even with the dozens of observing systems now used, there are still limited observational data to produce space weather forecasts.
- Research observing systems provide important data used to advance science; many of those also provide timely data and are used to support operational space weather services.
  - Several NASA heliophysics research missions will reach end-of-life within the next 10 years.
- Several NOAA and DoD space-based operational systems are scheduled to be replaced over the next 10 years subject to available funding.
- While NOAA, DoD and U.S. Geological Survey (USGS) ground-based systems are an important contributor to the space weather mission, sparse coverage limits their utility in meeting operational requirements.
- A number of foreign space-based and ground-based capabilities are used to help meet U.S. operational space weather needs.
  - More are available and provide the potential for future use.
  - While foreign data sources can provide additional capability, the economic and national security interests of the United States

dictate that the Nation not rely exclusively on foreign assets to conduct the critical space weather mission.

- Most unexploited data sources (foreign and domestic) are not currently used due to lack of reliable or timely access, excessive expense, policy/security restrictions, or other practical reasons. Also, these data sources offer secondary capabilities that cannot replace key, primary systems. Nevertheless, many offer added value that could incrementally improve forecasting, and should be used when feasible and cost-effective.

- While space-based and ground-based observing systems are critical components needed to meet operational requirements, they are inextricably linked to other parts of the space weather architecture (such as models and other space weather forecasting capabilities), and thus should not be considered alone when assessing our ability to meet requirements.

Table ES-1. Requirements Satisfaction by Phenomena

(A) Worst Case

	Nowcasts (Current Conditions)			Short-term Forecasts (minutes to hours)			Long-term Forecasts (hours to days)		
Timeline (years)	0-3	4-7	8-12	0-3	4-7	8-12	0-3	4-7	8-12
Geomagnetic Storms	G	G	G	Y(3)	O	O	Y(4)	O	O
Radio Blackouts	G	G	G	O	O	O	Y	Y	Y
Solar Radiation Storms	G	G	G	Y(2)	R	R	O	O	O
Ionospheric Storms	Y(1)	O	O	Y(3)	O	O	Y(4)	R	R
Atmospheric Drag	Y(1)	O	O	Y(3)	O	O	Y(4)	R	R

Substantial degradation over time if systems aren't sustained or replaced

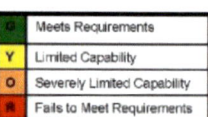

G	Meets Requirements
Y	Limited Capability
O	Severely Limited Capability
R	Fails to Meet Requirements

(1) Reduced DMSP coverage from two to one orbits
(2) Loss of relativistic electron data SOHO.
(3) Uncertainty of solar wind data from L1 to replace ACE.
(4) Uncertainty of getting a space-based coronagraph to replace SOHO and STEREO data.

Table ES-1. (Continued)
(B) Best Case

Timeline (years)	Nowcasts (Current Conditions)			Short-term Forecasts (minutes to hours)			Long-term Forecasts (hours to days)		
	0-3	4-7	8-12	0-3	4-7	8-12	0-3	4-7	8-12
Geomagnetic Storms	G	G	G	Y(3)	Y(5)	G	Y(4)	Y	Y
Radio Blackouts	G	G	G	O	O	O	Y	Y	Y
Solar Radiation Storms	G	G	G	Y(2)	Y	Y	O	O	O
Ionospheric Storms	Y(1)	Y	Y	Y(3)	Y	Y	Y(4)	Y	Y
Atmospheric Drag	Y	O	O	Y(3)	Y	Y	Y(4)	Y	Y

Requirements Satisfaction maintained or Improved if key systems are sustained or replaced

G	Meets Requirements
Y	Limited Capability
O	Severely Limited Capability
■	Fails to Meet Requirements

(1) COSMIC-2 deployed
(2) Relativistic electron data from SOHO are obtained.
(3) Solar wind data from L1 to replace ACE is obtained.
(4) Space-base coronagraphs on SOHO and STEREO are replaced.
(5) Advanced plasma sensor on DSCOVR follow-on obtained.

Observing systems referenced above:
ACE: Advanced Composition Explorer
COSMIC-2: Constellation Observing System for Meteorology, Ionosphere, and Climate - 2
DMSP: Defense Meteorological Satellite Program
DSCOVR: Deep Space Climate Observatory
SOHO: Solar and Heliospheric Observatory
STEREO: Solar TErrestrial RElations Observatory
* Observing systems referenced above:
COSMIC-2: Constellation Observing System for Meteorology, Ionosphere, and Climate - 2 GOES-R: Geostationary Operational Environmental Satellites - R
SEON: Solar Electro-Optical Network
SSAEM: Space Situational Awareness Environmental Monitoring

1. INTRODUCTION

On August 2, 2011, the Joint Action Group for Space Environmental Gap Analysis (JAG/SEGA) presented a briefing, titled *Space Environmental Gap Analysis*, to the National Space Weather Program Council (NSWPC), with staff members of the Office of Science and Technology Policy (OSTP) and the

Office of Management and Budget (OMB) in the Executive Office of the President present as observers. The purpose of the briefing was to present interagency findings regarding space weather observing systems, including an assessment of the current systems and requirements for the next 10 years. This report formally documents the findings, including additional explanatory information, by directly capturing key text and graphics from the briefing. This introductory section provides background information, the objective and scope for the assessment, and the methodology of how the assessment was conducted (including JAG/SEGA participants). Subsequent sections provide additional context and supporting material, to include: a discussion of the relevance and requirements; a summary and description of space weather observing systems; a discussion of the analysis, to include the methodological framework and results; and a summary of the findings from the JAG/SEGA and of the NSWPC.

1.1. Background

The 2010 National Aeronautics and Space Administration (NASA) Authorization Act, Section 809 (see Appendix 1) acknowledges:

- the threat to modern systems posed by space weather events;
- the potential for "significant societal, economic, national security, and health impacts" due to space weather disruptions of electrical power, satellite operations, airline communications, and position, navigation and timing systems; and
- the key role played by ground-based and space-based space weather observing systems in predicting space weather events.

In addition, the Act directed OSTP to submit a report to the appropriate Congressional committees that details the following:

- "Current data sources, both space- and ground-based, that are necessary for space weather forecasting."
- "Space- and ground-based systems that will be required to gather data necessary for space weather forecasting for the next 10 years."

In response to Congressional guidance, OSTP asked the Office of the Federal Coordinator for Meteorological Services and Supporting Research

(OFCM) on April 8, 2011, to lead the coordination of a new interagency assessment, through the NSWPC, and to provide to OSTP a report to address the Act's requirements. To conduct the assessment, the NSWPC established the JAG/SEGA on April 28, 2011.

1.2. Objective

The primary objective of this assessment was to support OSTP in responding to Congressional guidance put forth in the 2010 NASA Authorization Act. As such, the specific objectives of this report are:

- Detail the current data sources, both space- and ground-based, that are necessary for space weather forecasting.
- Detail the space- and ground-based systems that will be required to gather data necessary for space weather forecasting for the next 10 years.

To meet these objectives, the NSWPC was tasked with the following deliverables to OSTP:

- Provide an interim status briefing by end of July 2011.
- Provide a Report by end of September 2011.

1.3. Scope

In defining the scope of this assessment, the JAG/SEGA used the following determinations to guide the methodology and completion of the assessment:

- **Requirements**: Proceed from currently documented observing requirements for operational space weather services.
  - Given the short timeline required for this assessment, and the fact that the observing requirements from Department of Defense (DoD), Department of Commerce (DOC) National Oceanic and Atmospheric Administration (NOAA), and NASA were recently validated (see section 2.4), a formal revalidation of these

requirements was not considered to be needed to conduct this assessment.

- The scope was limited to observing requirements and systems necessary to drive operational forecasts and services. Requirements for purely research purposes without operational applications were not considered within the scope of the study, noting that the ongoing National Research Council (NRC) Decadal Survey on Solar and Space Science is assessing research plans and needs.
- Requirements for observations needed to support space missions beyond Earth geosynchronous orbit (lunar, interplanetary, etc.) were also considered.
- **Observing Systems**: Use existing agency requirements, programs, initiatives, and plans for observing and forecasting systems.
  - Only existing or planned systems were considered. Consideration of potential new systems beyond those already planned was considered to be out of scope.
  - Systems included in the assessment were operational systems and research platforms that are (or can be) leveraged for operational use. Research systems that are not conducive for operational use were not within the scope of the study.
  - International capabilities that can be leveraged to support U.S. operational needs were also considered.

1.4. Methodology

Leveraging the OFCM interagency coordinating infrastructure, the NSWPC established the Joint Action Group for Space Environmental Gap Analysis (JAG/SEGA) to perform an assessment of existing and planned space weather observing systems (see Appendix 2). The JAG/SEGA included representatives from the array of U.S. Government space weather stakeholders, with 25 participants from 15 organizations. As the providers of the Nation's two primary operational space weather analysis and forecasting centers, leaders from the U.S. Air Force (USAF) and the NOAA volunteered to serve as co-chairs for the JAG. The other JAG members represented the major stakeholder organizations in the national space weather enterprise, and made significant contributions to the assessment. Table 1 lists the key members of the JAG and other participating organizations; the full list of individual JAG members is contained in Appendix 2.

Table 1. JAG/SEGA Participants

JAG/SEGA Key Members and Participating Organizations	
Name (role)	Organization
Dr. Bill Denig *(Co-chair)*	NOAA National Environmental Satellite, Data, and Information Service (NESDIS)
Col John Egentowich *(Co-chair)*	Air Force Directorate of Weather (A3O-W)
Jerry Sanders *(Aurora Domain Lead)*	Air Force Weather Agency (AFWA)
Dr. Arik Posner *(Heliosphere Domain Lead)*	NASA HQ
Kelly Hand *(Ionosphere Domain Co-Lead)*	Air Force Space Command (AFSPC)/Aerospace Corp.
Dr. Therese Moretto Jorgensen *(Ionosphere Domain Co-Lead)*	National Science Foundation (NSF)
Dr. Michael Hesse *(Magnetosphere Domain Lead)*	NASA Goddard Space Flight Center (GSFC)
Bill Murtagh *(Solar Domain Lead)*	NOAA National Weather Service (NWS)
Clayton Coker *(Upper Atmos. Domain Lead)*	Naval Research Laboratory (NRL)
Michael Bonadonna *(Executive Secretary)* Other Participating Department of Energy (DOE) National Nuclear Security Admin. (NNSA)	Office of the Federal Coordinator for Meteorology (OFCM) Organizations Office of the Assistant Secretary of Defense for Networks and Information Integration [OASD(NII)]
Department of State (DOS)	
US Geological Survey (USGS)	AF Space & Missile Systems Center (SMC)

The methodology adopted by the JAG/SEGA was streamlined to focus on the specific goals set forth by Congress in the 2010 NASA Authorization Act, and to provide rapid results to meet the Act's timelines. The JAG collected, collated, and determined short-term and long-term space weather observing requirements needed to support operational space weather forecasting. A detailed description of the requirements is provided in Section 2.

In assessing the existing and planned space weather observing systems needed to meet these requirements, the JAG considered ground-based and space-based solutions specifically designed for operations, research systems that are capable of being exploited for operations, and other domestic or international solutions that could be leveraged for operations. A detailed description of these systems is provided in Section 3. Some additional information regarding international capabilities is included in the "Additional Notes" section below.

The JAG then used its compilation of the requirements, along with the existing and planned observing systems to be used to satisfy those requirements, to perform an analysis to determine key requirements shortfalls, or gaps ("gap analysis"). The methods used in performing the analysis, and well as the results of the analysis, are described in section 4. A summary of the key findings are then presented in section 5.

Additional Notes:

1. The JAG took a conservative approach with respect to funding of current and planned systems in order to define realistic "best case" and "worst case" scenarios for observing system availability. In this sense, the "best case" and "worst case" mean the following:

- "Best case" = all the identified key systems are funded and successfully deployed.
    - It does not mean that additional improved capabilities are fielded that are not already identified as a program, nor does it mean that new scientific breakthroughs are made.
- "Worst case" = none of the identified key systems are funded and successfully deployed.
    - It does not mean that other baseline observing capabilities and infrastructure are lost; those are assumed to continue as part of this scenario.

2. In conducting its analysis, the JAG took into consideration existing or planned and securely funded international efforts for space weather observations. In addition to those efforts, the JAG is aware of international organizations with space weather equities that could prove useful in the future in helping America meet its space weather observational requirements. Four of

these efforts are discussed briefly below. While these collaborations do not drive the key findings found in this report, they provide a foundation for increased, mutually beneficial efforts that might aide U.S. efforts to meet its space weather observational needs.

- The World Meteorological Organization (WMO) has launched an Interprogramme Coordination Team for Space Weather (ICT-SW). This team consists of representatives from approximately 20 nations and is co-chaired by the United States and China. The ICT-SW has completed an assessment of space weather observation parameters and is preparing a Statement of Guidance, an effort broadly parallel to this JAG, with a nominal delivery to WMO by the end of the year.
- The International Space Environment Service (ISES) is a permanent service supported by four different international organizations. With its current Director based in Ottawa, ISES operates 13 space weather regional warning centers around the globe providing global, standardized, and free exchange of space weather information as well as monthly reports summarizing the status of satellites in Earth orbit and in the interplanetary medium.
- The International Living Star (ILWS) program is a coordinating activity between NASA and partners from international space agencies. The ILWS mission is to stimulate, strengthen, and coordinate space research to understand the governing processes of the connected Sun-Earth System as an integrated entity. ILWS activities include the entire spectrum from space mission coordination as well as planning for data sharing for space weather forecasting and analysis purposes.
- US government technical agencies, including NASA, NOAA, NSF, and USGS also maintain a wide range of international collaborations in addition to those identified elsewhere in the text.

2. RELEVANCE, CONTEXT, AND REQUIREMENTS

A number of reports and assessments have documented the effects of space weather on activities, systems, and human health on the ground, in the air, and in space. Also, Congress acknowledged the importance of space weather's impacts on the Nation in its guidance to OSTP as part of the NASA Authorization Act of 2010. Therefore, only a brief reminder of the importance of space weather is given here to establish the broader context for the specific observing requirements that follow. A discussion of the manner in which requirements are defined is then provided, beginning with a description of how observing systems fit into the overall space weather context, followed by an explanation of how observing requirements are parsed across the relevant space environment domains, and concluding with a summary of where these requirements have been documented.

2.1. Relevance of Space Weather - Why It Is Important

National infrastructure and services are complex and interdependent; a major outage in any one area has a widespread impact. As noted in the 2008 NRC report, *Severe Space Weather Events*, "potential damage resulting from these critical dependencies can be minimized by having a robust capability to monitor, model, and predict what is happening in the space environment." Examples of key dependencies and impacts include:

- **Electric Power Grid:** Large-scale blackouts and permanent damage to transformers, with lengthy restoration periods.
- **Global Satellite Communications:** Widespread service disruptions, which can impact financial, telemedicine, government, and Internet services, among many others.
- **Global Positioning System (GPS) Positioning and Timing:** Degradations of military weapons accuracy, air traffic management, transportation, precision survey/construction, agriculture, energy exploration, ship navigation/commerce, financial transactions, and cell phone/broadband.
- **Satellites & Spacecraft:** Loss of satellites and capabilities, loss of space situational awareness (including detection of hostile actions), increased probability of satellite-debris collisions, degraded communications/navigation, and increased risk to astronaut safety.

For operators and decision makers to be able to take actions to mitigate these negative impacts, they must first have situational awareness of the space weather events that cause these impacts. Knowledge that a significant space weather event is occurring, as well as timely and accurate forecasts of the future state of the space environment, provides the means to take proactive measures to mitigate the impacts of these potentially damaging space weather events. It is this approach that led NOAA to develop Space Weather Scales for geomagnetic storms, solar radiation storms, and radio blackouts (see Appendix 3).

The impacts of space weather can have serious economic consequences. For example, geomagnetic storms during the 1990's knocked out several telecommunications satellites, which had to be replaced at a cost of about $200 million each. If another "once in a century" severe geomagnetic storm occurs (such as the 1859 "super storm"), the cost on the satellite industry alone could be approximately $50 - $100 billion. The potential consequences on the Nation's power grid are even higher, with potential costs of $1 - 2 trillion that could take up to a decade to completely repair.

(For above cost references, see: http://www.economics.noaa.gov/?goal= weather&file=events/space)

More detail on the importance of space weather impacts on society is provided in Appendix 4, which was previously published as part of the National Space Weather Program Strategic Plan (June 2010).

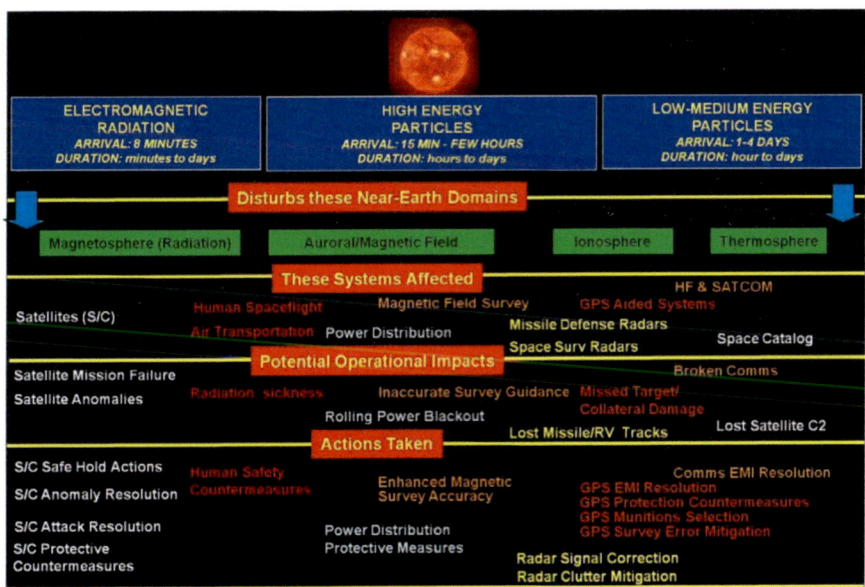

Figure 1. "Conditions-Systems-Impacts-Actions" Linkage.

Based on knowledge of how space environmental conditions can negatively impact certain systems, space-environmental monitoring and forecasting provides actionable information to operators and decision makers who can take actions to mitigate these risks and impacts. This linkage of space environmental conditions, systems, impacts, and actions is depicted in Figure 1. The figure illustrates how three space weather conditions (blue boxes) disturb four domains in the near-earth environment (green boxes). These disturb systems highlighted in the middle of the figure with potential impacts (in the same color) directly below each system. Finally, actions that can be taken to mitigate the impacts are shown (in the same color) on the lowest tier.

2.2. Space Weather Architecture

At a high level, the architecture for space weather observing and forecasting can be described in terms of three basic components, as depicted in Figure 2. The first component is the suite of space-based and ground-based observing systems that measure the space environment, which is the focus of the assessment detailed in this report. Measurements from these observing systems feed into the second component, which are the operational space weather centers composed primarily of the National Weather Service's Space Weather Prediction Center and the Air Force Weather Agency, as well as NASA's Space Weather Laboratory. At these centers, the measurements from all available sensors are processed, assimilated, and used as input to numerical prediction models to produce analyses (i.e., "nowcasts"), short-term forecasts (on a timescale of minutes to hours), and long-range forecasts (on a timescale of hours to days) of space weather events that are used to provide actionable products to operational users. In so doing, the analyses and forecasts of the space environment enable the centers to provide warnings and forecasts to operational users that take action to mitigate the space weather effects and risks described above.

There are several foundational building blocks that help support operational users. First, data assimilation techniques are used to ensure that data are properly incorporated for use in forecast models. Second, the science and technical know-how behind the models, the assimilation techniques, and other components of the process are continually updated and enhanced through a "research to operations" approach that is supported by government and university modeling centers (e.g., Community Coordinated Modeling Center, NSF Center for Integrated Space Weather Modeling, NRL), developmental test-beds, and prototyping/transition centers (e.g., AFWA, NOAA Space Weather Prediction Center (SWPC), Air Force Research Laboratory (AFRL) Space Weather Forecast Lab). Third, when combined with the underlying data networks and IT systems, the entire space weather analysis and forecasting *infrastructure* used by the centers is maintained to support the final component of the space weather architecture —the user community. Because all of these components are interdependent and linked, an assessment of the entire space weather architecture to meet current and future requirements must include an assessment of the analysis and forecast capabilities of the centers. The present assessment, however, is focused on the observing systems component.

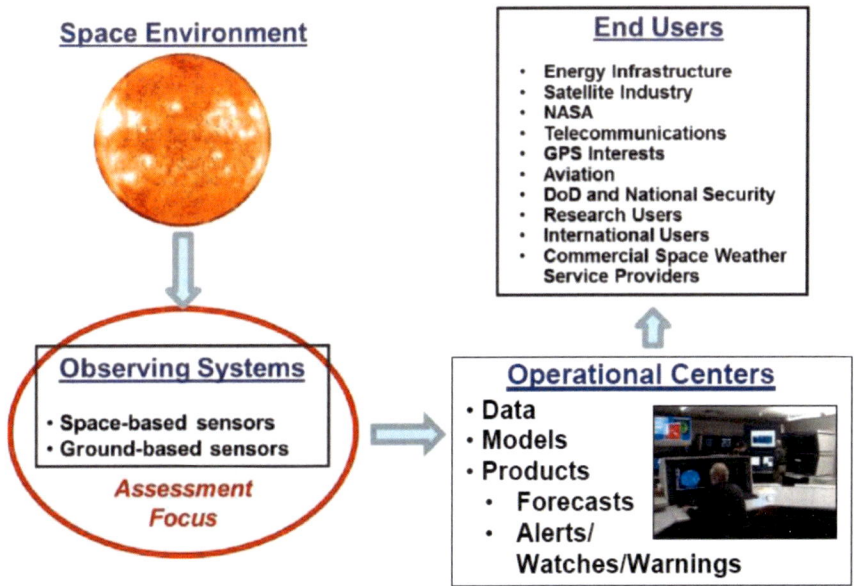

Figure 2. Space Weather Architecture.

2.3. Space Weather Domain Descriptions

As noted in the previous section, this assessment focuses on space weather *observing* requirements and capabilities and does not delve into the intricacies of the remaining parts of the space weather architecture, such as forecast models and customer products. In this context, an observational requirement is defined as a documented need for a *measurement* of a space environmental parameter, and is contingent on the "domain" of the space environment in which the parameter is measured. For this assessment, observing requirements are categorized within the following six domains of the space environment:

Sun/Solar, Heliosphere, Magnetosphere, Aurora, Ionosphere, and the Upper Atmosphere. As depicted in Figure 3, these domains span the space environment from the Sun to the Earth's atmosphere. Each domain has its own unique characteristics and importance to space weather, and is described in further detail below.

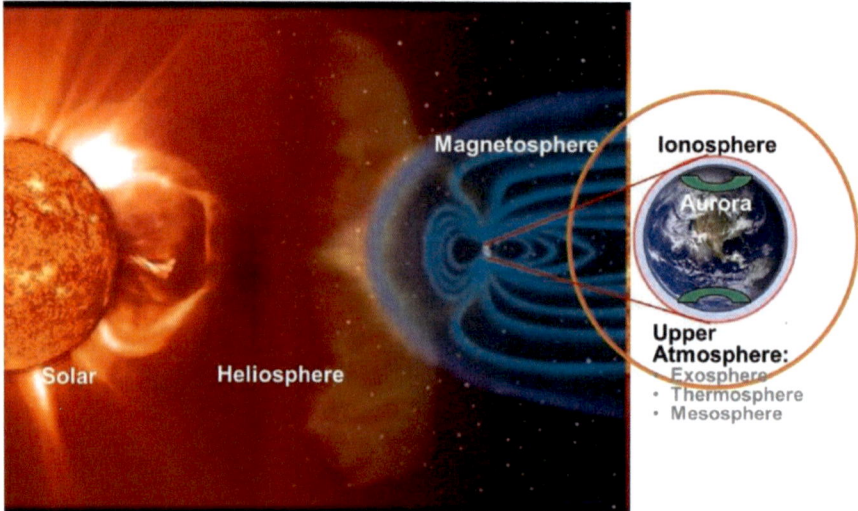

Figure 3. Space Weather Domains.

Solar: The Sun is the ultimate source of all space weather on or near the Earth. The solar domain consists of conditions near the surface, including the solar corona out to approximately 20 solar radii (Rs) and within the interior of the Sun, and is important to space weather in several ways. Monitoring conditions on the surface and in the interior of the Sun are used to detect the occurrence and precursors of solar flares. Prompt effects of solar flares at the Earth include increased ionospheric densities from energetic photons, mostly within the X-ray band, that ionize atmospheric gases. Flares are also indicative of major solar events that release vast amounts of solar gases in coronal mass ejections (CME), and energetic protons resulting in geomagnetic storms and polar-cap absorption events, respectively.

Heliosphere: The heliosphere is the immense magnetic bubble containing our solar system, solar wind (the plasma of charged particles coming out of the Sun), and the entire solar magnetic field, stretching out some 18 billion kilometers from the Sun. For space weather impacts, the area of most concern is with the inner heliosphere from within 1 Astronomical Unit (AU), approximately 150 million kilometers at the Earth location, to about 1.5 AU for Mars. It takes approximately 8 minutes for solar photons traveling at the speed of light to reach Earth, whereas it can take up to several days for the solar wind and intermittent solar gases emitted from the Sun in the form of CMEs to cover the same distance. Monitoring the heliosphere allows space weather operators to forecast whether and when a solar transient, such as a CME, might cause a magnetic storm on Earth. Included in the current assets available to forecasters is the Advanced Composition Explorer (ACE) satellite at the L1 Lagrangian point close to the Earth at approximately 240 Earth Radii (RE), approximately 1.5 million kilometers, along the Earth-Sun line. From this vantage point, operators can provide a short-term forecast, on the order of 45 minutes. Other assets monitor the inner heliosphere much closer to the Sun, thereby facilitating longer-term forecasts of up to several days.

Magnetosphere: The magnetosphere is the magnetic cavity surrounding the Earth, carved out of the passing solar wind by virtue of the Earth's magnetic field (or geomagnetic field), which prevents, or at least impedes, the direct entry of the solar wind plasma into the cavity. On the dayside extent (towards the Sun) of the magnetosphere, out to what is referred to as the magnetopause, is of order 8-10 RE. This dayside protective shield essentially blocks the solar wind and is highly responsive to changes in the solar wind speed and direction plus variations in the orientation of the interplanetary magnetic field (IMF) that is carried with the solar wind and can couple into the geomagnetic field near the magnetopause. Large solar wind impulses at the magnetopause can be monitored as magnetic field perturbations by satellites in geostationary orbit at approximately 7.7 RE and on the ground at magnetic observatories (such as those maintained by USGS). On the night side, the solar wind tends to drag out the geomagnetic field to distances of up to several hundred RE into what is referred to as the magnetotail. Magnetic reconnection between the IMF and geomagnetic field on both the dayside and night side can transfer enormous amounts of energy from the solar wind to the geospace environment. Geomagnetic storms occur when energy transferred from the solar wind is deposited in the magnetotail, sometimes building up to point whereby a fraction of the energy is dumped into the near-Earth space environment in the form of a magnetic substorm. Monitoring the magnetosphere in terms of the magnetic topology and energetic space particles allows operators to detect the occurrence of geomagnetic storms and to forecast the likelihood of resultant magnetic substorms.

Aurora: The aurora is a phenomenon associated with geomagnetic activity which occurs mainly at high latitudes; typical auroras appear in the thermosphere at approximately 100-250 km above the ground. The optical aurora is due to the collisional interaction between atmospheric gases, mostly neutrals, and precipitating energetic electrons and protons that stream along magnetic field lines from the more distant magnetosphere. The precipitating charged particles are typically of sufficient energy to collisionally ionize the atmospheric gases resulting in increased electron densities within ionospheric E and F layers that can be disruptive to radiowave propagation for communications and navigation. During geomagnetic storm periods (typically days), the occurrence of geomagnetic substorms (typically hours in duration) can lead to dramatic increases and changes in the electron density profile within the auroral zone as well as spectacular auroral displays that, at times, can be seen overhead at lower latitudes in response to increased geomagnetic activity. Energy inputs from precipitating charged particles and incoming Alfven waves can lead to large spatial and temporal variations in electron density that causes, by way of one example, radar auroral clutter that can compromise the performance of military early warning radars. Energy inputs during geomagnetic storms can also cause increased satellite drag due to atmospheric heating and the resultant outward expansion (diffusion) of the upper atmosphere.

Ionosphere: The ionosphere is the region of the Earth's upper atmosphere containing a small percentage of free electrons and ions produced by photoionization of the constituents of the atmosphere by solar ultraviolet radiation at very short wavelengths (< 0.1 microns). While the fractional percentages of electrons and ions are small, the morphology of the ionosphere has profound effects on radio-wave propagation. Airline operations, particularly at high geographic latitudes, are critically dependent on the steady-state ionospheric structure for high-frequency (HF) communications; the occurrence of D-region absorption events (see Appendix 5), also referred to as polar-cap absorption events, is particularly troublesome. Radio propagation delay through the ionosphere impacts the accuracy of navigation, radar, and geolocation systems. Ionospheric scintillation resulting from small-scale variations in density can degrade the performance of communications and navigation systems. Low-latitude scintillation results from unstable height variations in density that can occur in the post-sunset low-latitude ionosphere. Scintillation can also occur at higher latitudes in the auroral zones (see radar auroral clutter in the Aurora domain discussion) due to particle precipitation and within the polar cap due to density variation in polar-cap patches. The ionosphere is a complex region of space that is intimately coupled to both the magnetosphere and atmosphere. While numerous operational assets are currently available to monitor the ionosphere, the complexity and temporal variability of this domain limits the utility of any single approach. Instead, the ensemble of data available from different techniques offers the best opportunity to fully specify and possibly forecast this domain.

Upper Atmosphere: The upper atmosphere is categorized as that part of the Earth's atmosphere above the stratosphere, made up of three distinct layers: the mesosphere (approximately 50-90 km), the thermosphere (approximately 90-600 km), and the exosphere (approximately 600- 100,000 km). While the upper atmosphere is not nearly as complex as the ionosphere, the tools available for monitoring this domain are limited. Specifying this domain is important for calculating atmospheric drag effects on space systems including functioning satellites, space debris, and re-entry vehicles. Quasi steady-state specifications of the upper atmosphere can be effectively modeled for atmospheric drag using, for example, diurnal and longer term solar-cycle variations in solar heating. Less quantified are the variations in the heat flux from the magnetosphere during geomagnetic storms that can lead to dramatic changes in localized atmospheric drag. Specifying this domain is also important as it impacts the ionosphere in multiple ways. Variations in the thermospheric winds impact plasma redistribution in the ionosphere and are not effectively modeled.

2.4. Basis of Requirements

To adequately specify each of the six space weather domains previously discussed, several environmental parameters (i.e., specific observational requirements) must be measured. Table 2 lists the various environmental parameters needed to specify each domain. Specific environmental parameter measurements are used by the operations centers to provide nowcasts and forecasts of space weather. More details for each observed parameter, along with a description of why each is important, are presented in Appendix 5.

In analyzing the operational observing requirements, the JAG/SEGA made use of the most recent requirements documents from the two Federal departments that run the U.S. operational space weather centers, namely the DOC and DoD, as well as from NASA that operates research satellites (many of which are leveraged for operations) and their Space Weather Laboratory. The requirements used in this assessment are formalized in the following documents:

- NOAA Consolidated Operations Requirements List, 2011 (DOC).
- NOAA Program Observation Requirements Document – Space Weather Program, 2009 (DOC).
- Air Force Weather Space Weather Implementation Plan, Oct 2010 (DoD).
- Initial Capabilities Document for Meteorological and Oceanographic Environment, 2009 (DoD).
- Integrated Space Weather Analysis System Data Requirements, 2011 (NASA).
- Space Radiation Analysis Group Requirements, 2011 (NASA).
- Four-Dimensional Weather Functional Requirements for NexGen Air Traffic Management, 2008 (Joint Planning Development Office Weather Functional Requirements Study Group).

Table 2. Observing Requirements by Space Weather Domain

Solar	Heliosphere	Magnetosphere	Aurora	Ionosphere	Upper Atmosphere
Solar EUV &UV Flux	Solar Wind: 3D Mag. Field Components	Energetic Ions and Protons: Energy & Flux	Auroral Boundaries (Equatorial and Polar)	Ionospheric Scintillation: Phase and Amplitude	Mesospheric Temperature
Solar EUV and UV Imagery	Solar Wind Plasma Components: Composition, Density and Temperature	Medium Charged Particles: Total Flux and Energy	Auroral Energy Deposition	Plasma Fluctuations	Mesospheric Wind Speed and Direction
Solar Magnetic Field	Solar Wind: Speed and Direction (3D Plasma Velocity Components	Trapped Particles: Protons, Electrons, Waves	Auroral Emissions & Imagery: UV, Visible and IR	Plasma Temperature: Te & Ti Plasma Temps	Neutral Winds (Speed & Direction)
Solar Radio Emissions: (Total and spectral flux)	Sun-Earth line Heliospheric Imagery	Supra-thermal through Auroral Energy Particles: Diff. Dir., Energy, Flux	Precipitating Particles: Electrons; 20eV-1KeV; 1KeV-50KeV	Ionospheric Characterizations: Layer Height & Freq.	Neutral Density, Composition, and Temperature
Solar Radio Burst: (Location, Type, Polarization)	Off-angle Heliospheric Imagery	Magnetic Field Strength and Direction		Energetic Ions 1-500MeV	Neutral Density Profile
Solar Imagery IR and Optical	Solar Wind Radio Emissions	Earth Surface Geomagnetic Fields		Total Electron Content	
Solar Coronagraph	Relativistic Electrons			Electric Field	
Solar X-Ray Flux (total and discrete Freq.)	Solar High Energy Protons and Cosmic Rays			D Region Absorption	
Solar X-Ray Imagery	Off-angle Solar Wind In Situ Parameters			Electron Density Profile: Density, Features, Composition	
Off-angle Solar Imagery					
Helio-seismology					

3. OBSERVING SYSTEMS FOR OPERATIONAL SUPPORT

There are several parallels between traditional atmospheric weather observing that is needed for forecasting, and the similar processes used for space weather. First, some observations are best taken remotely while others must be taken in situ to be useful. Second, both space-based and ground-based sensors are needed to measure various key environmental parameters. Third, space-based sensors are needed in different orbits to meet operational and research needs.

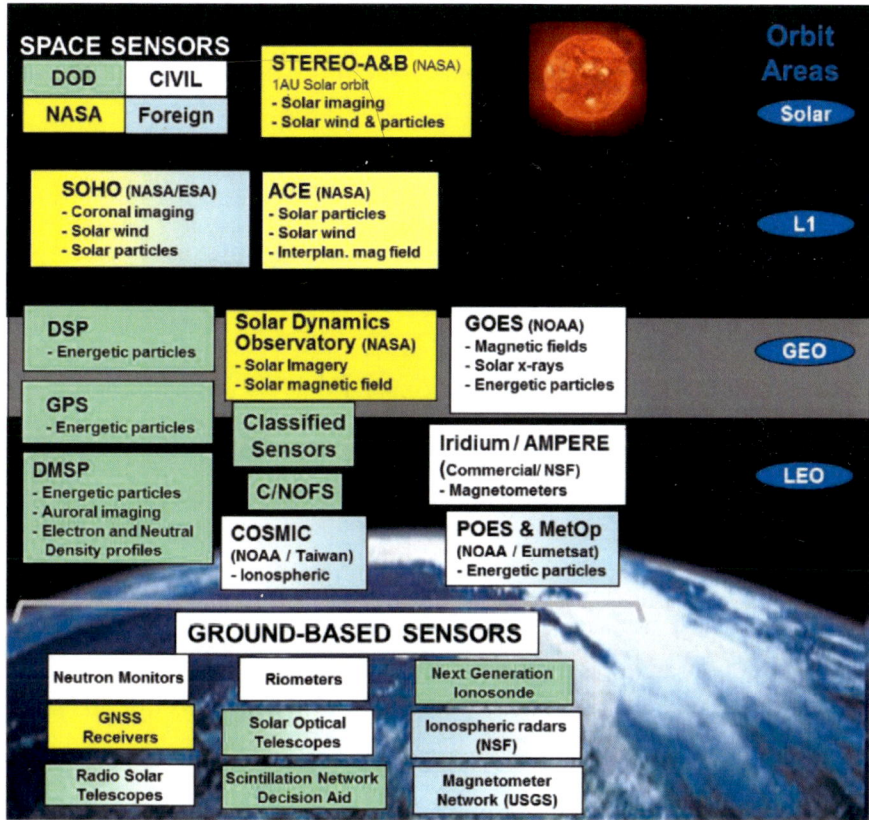

Figure 4. Space Weather Observing Systems.

One notable difference between these two environments is the density of observational data associated with each environment—the volume of insterstellar space is many orders of magnitude greater than the volume in which terrestrial weather conditions exist. Also, the number, variety, and coverage from space weather observing systems are small compared to atmospheric observing systems. While this results in limited observational data to produce space weather forecasts, the current suite of space weather observing systems, depicted in Figure 4, still provides significant capabilities in meeting many operational requirements.

In the subsections that follow, each observing system considered in this assessment is described; also, systems are grouped as either a ground-based system or space-based system. The system descriptions are grouped into three subsections, according to the following structure:

- Existing systems currently used for operations.
- Existing systems not currently used for operations (but could be with additional effort).
- Future/planned systems to replace/upgrade existing systems.

3.1. Existing Systems Currently Used for Operations

Ground-Based Systems

Digital Ionosonde Sounding System (DISS): Originally fielded by the USAF in the early 1990's, DISS was comprised of 20 unmanned automated sites strategically positioned to support USAF operations. DISS provides all standard ionosonde parameters, and data are retrieved in near-real-time for use in ionospheric models. DISS will be fully decommissioned by 2012 and replaced by NEXION. Figure 5 depicts the locations of DISS and other ionospheric sensors.

Global Oscillation Network Group (GONG): The GONG is a community-based program to conduct a detailed study of solar internal structure and dynamics using helioseismology. To exploit this new technique, GONG has developed a six-station network of extremely sensitive and stable velocity imagers located around the Earth to obtain nearly continuous observations of the Sun's "five-minute" oscillations, or pulsations. GONG is supported by the NSF National Solar Observatory and is expected to operate through 2022, subject to the outcome of the NSF Astronomy Division's current Portfolio Review process. GONG capabilities will be enhanced to include solar H-alpha observations in support of USAF needs during the ISOON development and deployment. See Figure 6 below for current GONG locations.

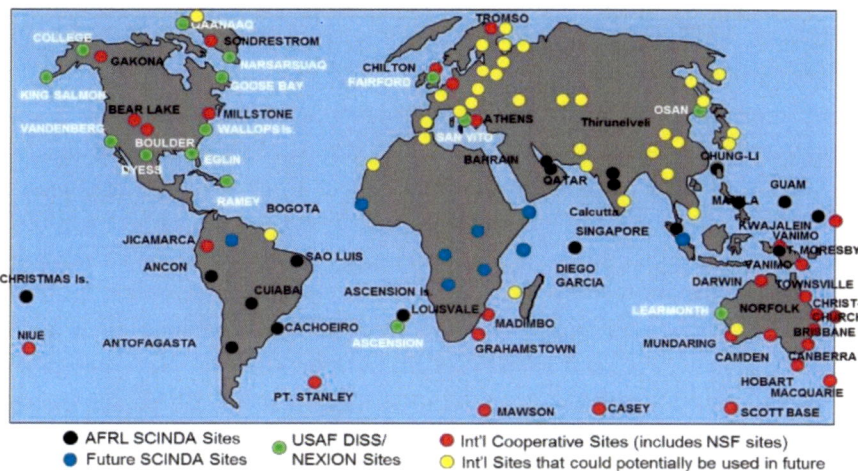

AFRL SCINDA Sites   USAF DISS/   Int'l Cooperative Sites (includes NSF sites)
Future SCINDA Sites   NEXION Sites   Int'l Sites that could potentially be used in future

Figure 5. Current ground-based Ionospheric Sensors.

Global Positioning System (GPS) Receivers: The superb accuracy of the GPS can be used to derive various ionospheric parameters, including Total Electron Content (TEC), Electron Density Profiles (EDP), and L-band scintillation. Within NOAA, the National Geodetic Survey (NGS) acquires GPS receiver data from approximately 1800 sites mostly within CONUS as part of the Continuously Operating Reference Stations (CORS) program. The CORS data are provided to the SWPC and assimilated into the US-TEC model. For DoD space weather operations, AFWA acquires globally-distributed GPS receiver data from the NASA Jet Propulsion Laboratory (JPL) TEC network. NASA uses the GPS data and information acquired from the Space Weather Application Center – Ionosphere (SWACI) operated by the German Aerospace Center. The increasing proliferation of ground receivers for GPS, as well as for other Global Navigation Satellite System (GNSS) programs, makes the use of these data attractive for space weather operations, although current sources are limited to land-based locations. Space-based GPS occultation sensors within the COSMIC and C/NOFS programs (discussed below) also make use of the GNSS network for space weather.

International Ionosondes: The U.S. space weather centers routinely access data from ionosondes operated by foreign agencies and organization to augment existing U.S. networks. The NOAA National Geophysical Data Center acquires international ionosonde data in near-real-time and provides these data to the operational centers. See Figure 5 above for locations of currently used sites, as well as potential new sites.

Neutron Monitors: The neutron monitor operated at Thule Air Base in Greenland provides real-time observations used to determine cosmic ray flux on the Earth's atmosphere. Galactic cosmic rays can be hazardous to people in space, on aircraft and on the ground, depending on the intensity. Solar cosmic rays can also be detected by the neutron monitors. Neutron Monitor data are the means to detect ground-level events. Data from several other neutron monitors are available through the European Space Agency (ESA) and other sources.

Next Generation Ionosonde (NEXION): Air Force Weather is currently fielding NEXION, a new digital solid-state sensor technology at up to 30 locations within the U.S. Air Force Ionospheric Data Network. These unmanned sensors provide near-real-time data to drive USAF ionospheric models for operational support. NEXION is expected to reach full operational capability in 2017 and remain in service well into the future. See Figure 5 for known NEXION locations.

Penticton Solar Radio Telescope: The Solar Radio Monitoring Program is a service operated jointly by National Research Council Canada and the Canadian Space Agency. Its function is to provide current and archival values of the 10.7cm Solar Flux solar activity index, which is a proxy indicator for the Extreme Ultraviolet (EUV) radiation striking the Earth's upper atmosphere giving rise to the ionosphere. The long uninterrupted history of 10.7cm flux measurements provides vital input for many ionospheric applications. Also, monthly Penticton 10.7 cm Radio Flux values are a primary input for measuring solar cycle progression.

Riometers: These sensors are used to measure the relative ionospheric opacity for radio signals and provide reliable information on the presence and density to the D-region of the ionosphere. Real-time riometer data are collected from Thule Air Base in Greenland and used by the operational space weather centers. Several other riometers are available but not routinely used.

Scintillation Network Decision Aid (SCINDA): SCINDA is a system designed to specify ionospheric scintillation in real time. Timely location of outage regions enable DoD users to effectively use satellite communication, navigation, or surveillance assets to modify mission plans and prevent errors as scintillation warnings become available. Specialized ground-based Ultra High Frequency (UHF) and L-Band receivers, monitoring signals from geosynchronous communication satellites, are used to measure scintillation intensities and zonal drift velocities. Data from the SCINDA sites are restricted for DoD use.

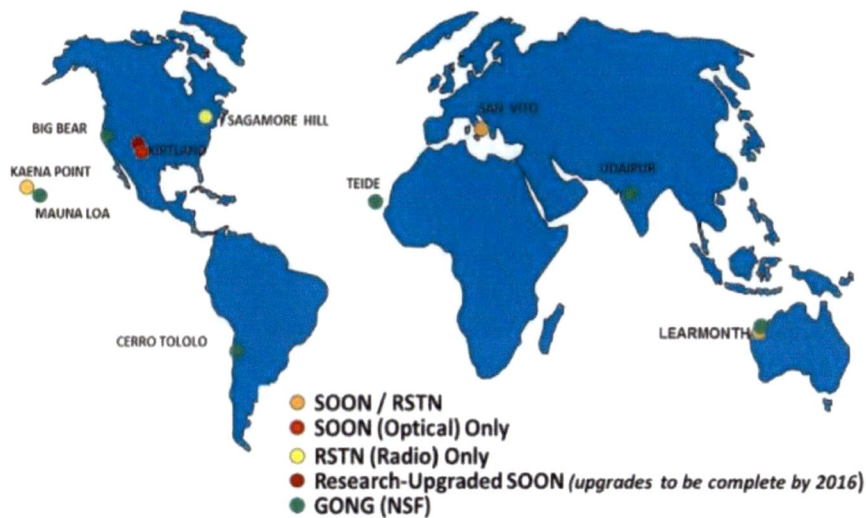

Figure 6. Ground-based Solar Telescopes.

Solar Electro-Optical Network (SEON): Since the 1960's, the USAF has operated solar optical and radio telescopes to support various missions affected by space weather.

The current SEON network provides 24x7 solar "patrol" which combines Hydrogen-alpha optical observations from the Solar Optical Observing Network (SOON), with a wide spectrum of solar radio emissions from the Radio Solar Telescope Network (RSTN). Continuing upgrades to SEON and its individual telescopes and components will keep the network services operating for the foreseeable future. See Figure 6 for SOON and RSTN locations.

USGS Magnetometers: The USGS owns and operates a network of 14 real-time magnetometers in the northern hemisphere across North America and the Pacific Ocean.

Data from these sensors are used for a wide variety of purposes, including monitoring of changes in the Earth's magnetic field, electromagnetic conditions in the ionosphere, and density and height of the atmosphere, which affects Low Earth-Orbit (LEO) satellites.

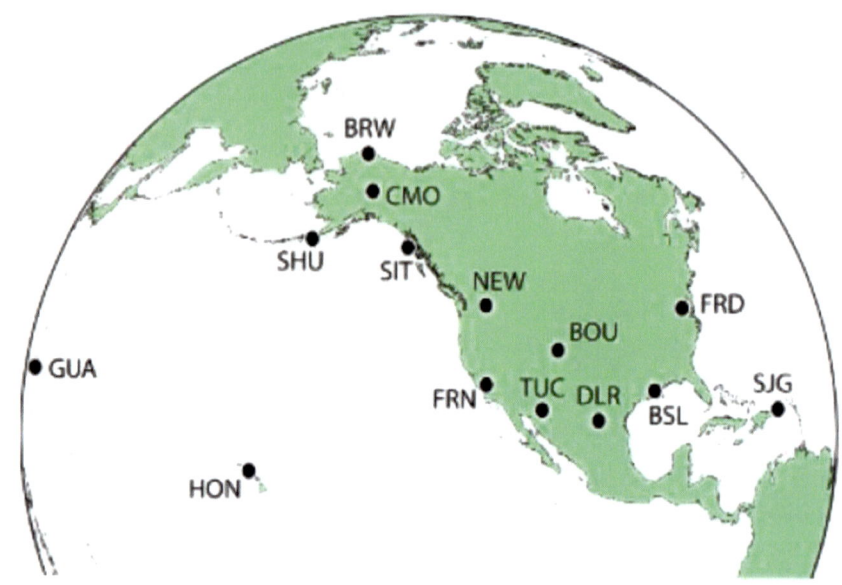

Figure 7. USGS Magnetometers.

Space-Based Systems

Advance Composition Explorer (ACE): Launched by NASA in 1997, ACE provides real-time scientific measurements of the solar wind from the Earth-Sun L1 point, located approximately 0.99 AU from the Sun and 1 million miles from Earth. It provides measurements of the interplanetary magnetic field, solar wind composition, speed, density, pressure and temperature. ACE plasma measurements can be severely degraded during solar radiation storms. ACE is roughly 10 years past its mission design life, but NASA plans to continue operating the mission through 2014 and may continue to operate it until 2020 subject to NASA funding and spacecraft health.

Communication and Navigation Outage Forecast System (C/NOFS): C/NOFS is an AFRL Advance Concept Development Test-bed mission composed of one small spacecraft in low inclination LEO, and associated ground systems. Launched in 2008, it provides data for quasi-operational and research use including ionospheric plasma fluctuations, ion velocity, in situ electric field, neutral wind parameters, electron density profiles, and many other parameters. C/NOFS mission end of life (EOL) is 2012 unless continuation funding is provided.

Constellation Observing System for Meteorology, Ionosphere & Climate (COSMIC): Taiwan's Formosa Satellite Mission #3, also known as COSMIC, uses the GPS radio occultation method for research and operational meteorological and ionospheric data. It provides cost effective measurements of atmospheric vertical temperature, moisture, and electron density profiles. COSMIC is a joint mission between Taiwan and the United States that is sponsored by NASA, NOAA, NSF, the Air Force Office of Scientific Research, the Office of Naval Research, and the Space and Missile Systems Center. COSMIC includes six microsatellites in LEO and associated ground systems. COSMIC EOL is expected in 2012.

Defense Meteorological Satellite Program (DMSP): DMSP has provided atmospheric and space environmental data for the DoD since the 1960's. The current DMSP spacecraft in sun-synchronous LEO provide fairly low latency (approximately 105 minutes) data including UV measurements of the ionosphere, auroral boundary and particle detection, in situ magnetic field, and other space weather parameters. The DMSP mission and observations should be available through 2025.

Geostationary Operational Environmental Satellite (GOES): The current series of NOAA's GOES is comprised of the three spacecraft (GOES-N, -O, and -P) and associated ground systems The space environmental sensors on GOES-NOP include a solar X-ray imager, X-ray flux monitor, energetic particle monitors, and a magnetometer. Data are provided to the operational centers in real time, which provides crucial data for the onset of solar radiation storms and radio blackouts. GOES-NOP EOL is approximately 2020.

Los Alamos National Laboratory (LANL) Geosynchronous Earth-Orbit (GEO): DOE's LANL provides a variety of space environmental in situ measures from geostationary platforms. These data include solar high energy proton and cosmic ray fluxes, medium and low energy charged particle data, and trapped radiation (protons and electrons). These data are used by the DoD for space weather analysis and monitoring and should be available through 2022 and beyond. At present, these data are not available for operational space weather outside of the DoD.

MetOp: MetOp is the polar-orbiting meteorological satellite system operated by the European Organisation for the Exploitation of Meteorological Satellites (EUMETSAT). The MetOp instrument complement includes a Space Environment Monitor 2 (SEM-2), identical to the SEM-2 particle sensors on POES (see below). Currently the MetOp-A satellite, launched in 2006, provides space environmental data in the mid-morning sun-synchronous circular polar orbit at approximately 840 km altitude. Overall, the MetOp A/B/C satellites will provide operational data through approximately 2021.

Polar Orbiting Environmental Satellite (POES): NOAA's POES satellites have provided continuous space environmental data from a LEO sun-synchronous orbit since 1978. The current series of POES spacecraft includes a SEM-2 package. Space environmental data are currently received from 5 POES spacecraft, although only the POES NOAA-19 satellite, launched in 2009, is considered operational. The POES series will end after NOAA-19, nominally in 2012. Although NOAA will provide continued meteorological satellite observations after POES, no SEM-like instrument is planned for the follow-on Joint Polar Satellite System (JPSS) spacecraft. After the NPOES restructuring in 2010, it was assumed that a DoD satellite with an AM orbit would provide a space environment monitoring package. Indeed, both the DMSP-19 and later, the DMSP-20 satellite will each include space environment measuring payloads in the early morning orbit. These measurements will continue until the end of life of the final satellite, DMSP-20, in the 2025-timeframe. In the mid-morning orbit, DMSP-18 will include the same payloads until it reaches end of life in the 2016-timeframe. Historically, these mid-morning observations are more consistently useful for taking these types of measurements. Therefore following the end of life for DMSP-18, the planned COSMIC-2 mission will be a key contributor to the collection of space environment measurements.

Solar Dynamics Observatory (SDO): The SDO was the first mission launched as a part of NASA's Living With a Star (LWS) Program, an initiative designed to understand the causes of solar variability and its impacts on Earth. Launched in 2010 into geostationary orbit, it provides high resolution spatial, spectral, and temporal observations of the Sun. In addition to providing science data sets to the research community, the SDO ground system provides a subset of data for real-time operational purposes. SDO's prime mission lasts until 2015. Extended operations are subject to NASA approval.

Solar and Heliospheric Observatory (SOHO): In 1995, NASA and the ESA launched SOHO to the L1 point to begin a two-year mission of scientific discovery. Some 16 years later, SOHO continues to provide critical solar and heliospheric observations, including the only space-based solar coronograph on the Sun-Earth line in operation today. Along with its other observations, this makes SOHO an important tool for space weather observation and forecasts. Extended mission operations are funded through 2014.

Solar Terrestrial Relations Observatory (STEREO): NASA's twin STEREO spacecraft were launched into heliocentric orbits at approximately 1

AU and have drifted nearly 120 degrees ahead and behind the Earth. Launched in 2006, the STEREO spacecraft provide "off-angle" observations of the Earth-Sun line, allowing space scientists and space weather operators to have 3-dimensional views of coronal mass ejections as well as observations of the far side of the Sun. The STEREO mission EOL is 2014, but may be extended pending funding and spacecraft status.

3.2. Existing Systems Not Currently Used for Operations

Ground-Based Systems

Incoherent Scatter Radars: The NSF and a number of foreign and international organizations own and operate a variety of incoherent scatter radars that are primarily used for research studies and applications. They provide very accurate observations of the ionosphere and upper atmosphere, but only have limited regional coverage. A few of these systems currently have automatic and real-time data capabilities; with additional infrastructure upgrades they could be fully exploited for operations, should the value added be deemed worth the added cost.

Figure 8. NSF Incoherent Scatter Radar.

International Real-time Magnetic Observatory Network (INTERMAGNET): INTERMAGNET is a global network of observatories monitoring the Earth's magnetic field. The program exists to establish a global network of cooperating digital magnetic observatories, adopting modern standard specifications for measuring and recording equipment in order to facilitate data exchanges and the production of geomagnetic products. Currently 44 countries provide data from 118 geomagnetic observatories. Data from INTERMAGNET could substantially improve analysis of the global and regional geomagnetic field if adequate communications could be secured to retrieve the data in near real time. See Figure 9 for worldwide locations of current INTERMAGNET sites.

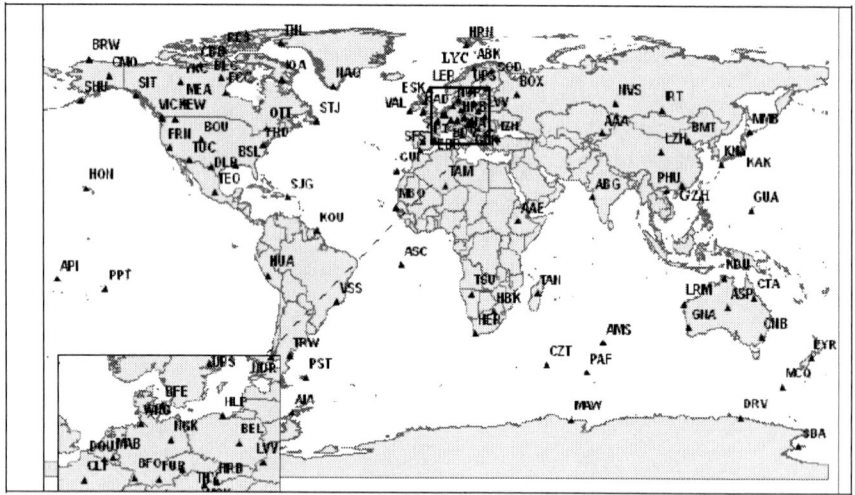

Figure 9. INTERMAGNET Sites.

Super Dual Auroral Radar Network (SuperDARN): SuperDARN consists of over 20 radars, operating on frequencies between 8 - 20 MHz and focused on the Earth's polar regions, which measure the position and velocity of charged particles in the Earth's ionosphere. Because the movements of these particles are tied to the movements of the Earth's magnetic field, which in turn extends into space, SuperDARN data provide scientists with information regarding the Earth's interaction with the space environment. SuperDARN is an international collaboration involving scientists and funding agencies from over a dozen countries. Although primarily a research tool, SuperDARN could be used for specific operational support if the operational space weather service providers developed and implemented data assimilation tools to exploit the data.

Space-Based Systems

Active Magnetosphere and Planetary Electrodynamics Response Experiment (AMPERE): The NSF has funded the AMPERE project to retrieve magnetometer data from the commercial Iridium communication satellites. These data could be processed to extract geomagnetic data and infer a wide variety of electrodynamic conditions on a global basis, but would first require the development of new data assimilation tools to exploit the data.

Los Alamos National Laboratory (LANL) GPS: LANL operates a number of particle and radiation sensors on the GPS constellation. However, use of these data has not been fully exploited outside of DOE. New data exploitation techniques would need to be developed in order to use these data for operations. (Note: see discussion of LANL GEO in previous section)

WIND: Launched by NASA in 1994, WIND collects data at the L1 point on solar wind speed, temperature and density, as well as the interplanetary magnetic field. Although the spacecraft is still functional, real-time data are not retrieved due to ground antenna costs and schedule conflicts.

3.3. Future/Planned Systems to Replace/Upgrade Existing Systems

GROUND-BASED SYSTEMS: Several ongoing upgrade programs (e.g., NEXION and SEON) are covered in section 3.1. No additional planned upgrade programs were identified or assessed as part of this study.

Space-Based Systems

Constellation Observing System for Meteorology, Ionosphere & Climate - 2 (COSMIC-2): COSMIC-2 will build upon the successful joint U.S.-Taiwan COSMIC mission due to be completed in 2012. COSMIC-2 will also use the GPS radio occultation method for research and operational meteorological and ionospheric data. Current plans call for launching as early as 2015 pending funding commitments. (Note: also see discussion of COSMIC in section 3.1)

Deep Space Climate Observatory (DSCOVR): DSCOVR is the planned near-term ACE replacement for providing in situ measurements of the solar wind and the interplanetary magnetic field at the L1 point. DSCOVR will provide critical data to meet all documented operational requirements and allow time for the development of a long-term national strategy for solar wind observations. NOAA will acquire DSCOVR from NASA for refurbishment, while the USAF will procure the launch vehicle.

DSCOVR Follow-on (DSCOVR-F/O): NOAA has been investigating the use of a commercial provider for solar wind data from the L1 point. This is envisioned as a possible long-term solution, after DSCOVR, for obtaining reliable, cost effective data. Some consideration is also being given to obtaining GPS occultation data in the post COSMIC-2 time frame.

Geostationary Operational Environmental Satellite - R (GOES-R): GOES-R is the follow-on program to NOAA's current GOES-NOP series of geostationary meteorological satellites. As with past GOES missions, the space environmental observations consist of in-situ measurements of energetic charged particles and local magnetic fields plus related solar observations. GOES-R solar measurements will continue NOAA's operational record of solar X-ray observations while shifting to the extreme ultraviolet band for solar imagery. The first launch of the GOES-R series satellite is scheduled for 2015.

Joint Polar Satellite System (JPSS): JPSS atmospheric soundings will be used to observe very high altitude measurements needed for the characterization of the neutral upper atmosphere. A key instrument for the JPSS is the Visible/Infrared Imager/Radiometer Suite (VIIRS). The VIIRS Day-Night Band (DNB) will provide space-based observations of the aurora under conditions of limited cloud cover and lighting (Sun and moonlight). Certain JPSS capabilities will also exist on the Suomi NPOESS Preparatory Project (NPP) satellite launched October 28, 2011.

Radiation Belt Storm Probes (RBSP): The RBSP is a NASA mission under the LWS program scheduled to launch a pair of identical spacecraft in low-inclination, Highly Elliptical Orbit (HEO) in 2012. The mission of RBSP is to gain scientific understanding of how populations of relativistic electrons and ions in space form or change in response to changes in solar activity and the solar wind. NASA plans to make these data available for operational use via a near-real-time beacon relay.

Space Environmental Nanosat Experiment (SENSE): SENSE consists of two cubesats being built by Boeing for SMC, with launch targeted for Fiscal Year (FY) 2013. Both satellites have a GPS receiver for ionospheric radio occultation. In addition to the GPS receiver, one also carries the Wind Ion Neutral Composition Suite (WINCS), an in situ sensor to measure solar wind, ions, neutral composition and ion drift. The other one will carry the Cubesat Tiny Ionospheric Photometer (CTIP), a UV photometer. The combination of sensors provides ionospheric specification at higher resolution than can be provided by radio occultation alone.

4. ANALYSIS

The JAG/SEGA, organized into six sub-groups for each of the space weather domains (see Appendix 2), performed a requirements analysis of space weather observing systems to respond to the Congressional direction posed in the 2010 NASA Authorization Act. Section 4.1 details the methodology used by the group to perform the analysis, while Section 4.2 details the results from the analysis.

4.1. Analysis Framework

The JAG/SEGA performed a detailed analysis for ground-based and space-based systems used to observe each of the six space weather domains. All current and planned observing systems used for operations were included in the assessment, as well as those not currently used but possibly useful for the future. Systems that are used exclusively for research and are not available for operations, for whatever reason, were excluded from the assessment. Each environmental parameter within the six space weather domains (see Appendix 5 for a list and description of the environmental parameters) was assessed against documented observing requirements.

While the analysis of the ability of current, planned, and potential systems to meet specific observing requirements was critical to the assessment, the JAG took an additional step to ensure that the end results were tied to real-world applications. The JAG mapped the observing parameters for each of the six domains to analysis and forecast products (nowcast, short-term forecast, and long-term forecast) for the five key space weather phenomena described below. The analysis included an assessment of the relative importance of each observed space environmental parameter for observing and forecasting the five space weather phenomena.

- **Geomagnetic Storms:**[†] A worldwide disturbance of the Earth's geomagnetic field resulting from increases in the solar wind pressure

---

[†] Phenomena included on NOAA's Space Weather Scales (see Appendix 3)

and interplanetary magnetic field at the dayside magnetopause. The occurrence of substorms within a geomagnetic storm period can negatively impact satellite operations, power systems, radio propagation, and navigation systems.

- **Radio Blackouts**:† Disturbances of the ionosphere caused by X-ray emissions from the Sun, which can negatively impact radio propagation and navigation systems.
- **Radiation Storms**:† The occurrence of elevated fluxes of charged particle radiation which can negatively impact satellite operations, radio propagation, navigation systems, and biological risks to humans in spacecraft or high-flying aircraft.
- **Ionospheric Storms**: Disturbances in the ionosphere caused by large increases in the fluxes of solar particles and electromagnetic radiation, often associated with the occurrence of geomagnetic storms. There is a strong coupling between the ionosphere and the magnetosphere which results in both regimes being disturbed concurrently. These disturbances can negatively impact radio communications as well as satellite navigation and communications systems.
- **Atmospheric Drag**: Collisions with diffuse air particles (altitudes typically < 2000 km) slowly act to slow down the spacecraft, leading it to gradually descend to lower altitudes where the drag continues to increase with increased atmospheric density. This is affected by space weather since the density of the air particles responds to solar activity, such as magnetic storms. Solar emissions cause the upper atmosphere to heat and expand, which in turn increases drag at a given altitude. This effect increases dramatically with high solar activity. If the increased solar activity triggers increased magnetic activity at the Earth, intense currents flowing through the upper atmosphere also contribute to increased heating and expansion of the upper atmosphere. Accurate analysis of atmospheric drag effects can reduce the error associated with determination of satellite orbital intersection with other satellites and space debris, reducing the need for expenditure of fuel for orbital maneuvers and thereby extending the mission life of the spacecraft.

4.2. Detailed Analysis Results by Space Environmental Domain

Using the methodology outlined in Section 4.1, the JAG/SEGA obtained detailed results within each of the six space weather domains. The specific details of this analysis are reported in Appendix 6, while the most significant results (i.e., the ones that most directly impact space weather operations) are provided below for each domain.

Sun/Solar: During the interval FY11-22, there is good coverage of the Sun provided by NOAA operational spacecraft, the various leveraged NASA assets, and the USAF SEON (which consists of the SOON and RSTN). During the operational transition from SOON to ISOON, additional ground-based optical coverage will be provided by the NSF GONG network. A high-risk capability over the next 10 years is the uncertain continuity of leveraged coronagraph observations provided by the NASA SOHO satellite which is currently operating in the "Bogart" mode, a reduced mode of operation at greatly reduced cost. In this mode, the critical white-light coronagraph observations from a Sun-Earth line view will continue, but from a satellite that is 14 years past its nominal mission lifetime. Additionally, while the NASA STEREO mission has demonstrated the utility of off-angle solar monitoring, the quality of off-angle coronagraph observations will diminish as the two satellites continue to depart from optimum position near the L4 and L5 Lagrangian locations and continue to separate in their heliocentric orbits.

Heliosphere: Reliable, operational observations of the solar wind and of the interplanetary magnetic field at L1 are perhaps the most important real-time data needed to create an effective level of operational space weather monitoring and forecasting. Currently, the availability of data for the heliospheric domain is heavily dependent on leveraged NASA assets. However, current real-time data provided by NASA research sensors are inadequate or may be interrupted during severe storm conditions, as demonstrated during the 2003 Halloween storms. Furthermore, the long-term continuity of NASA research-quality data is not assured through FY22. No current observational systems provide the capability to provide long-range forecasts of severe storms that have the potential to cause major impacts and drive most of the critical effects in geospace and on the surface. DSCOVR, along with the possibility of a potential commercial data buy solution, are planned and under consideration, respectively, as sequential follow-on replacements for ACE. While the current NOAA GOES-NOP satellites, which will transition to the GOES-R series after 2015, provide continuity of nowcasting, these satellites do not specifically address forecasting requirements. Current heliospheric imagery data provided by the Solar Mass Ejection Imager (SMEI) sensor on the Coriolis satellite, with its limited applicability to geomagnetic storm forecasting for Earth, will likewise be available only through the mid-term (4-7 years).

Magnetosphere: Key data for the magnetospheric domain are measurements of energetic charged particles. Measurements with thermal energies below 100 electron volts (eV) to 10's of keV are useful for surface charging assessments, while measurements of higher energy particles (in the MeV range) are used for high-latitude aviation interests, astronaut protection and to mitigate their deleterious effects on vehicle electronics. In addition, magnetic field measurements are important as they provide the means to assess the magnitude and progress of geomagnetic storms. A complete coverage of all relevant locations in geospace requires measurements along a variety of radial distances. Furthermore, it should be noted that data obtained from magnetospheric measurements alone strictly support only nowcasting and specifications, as well as post-event analyses. For forecasting purposes, solar wind measurements (e.g., from the L1 point or solar observations) are essential to augment even accurate specification of the current state of the magnetospheric environment. In the near-term (0-3 years) and midterm (4-7 years), the availability of leveraged energetic particle data from the pair of NASA RBSP spacecraft will provide good coverage of the magnetosphere during each 9-hour orbit period. Particle data from the NOAA GOES and POES spacecraft, along with the USAF DMSP satellites, provide supporting data, albeit with limited local time coverage. While there is the possibility to extend the lifetime of the RBSP, once this satellite mission ends the overall coverage of the magnetosphere will be substantially diminished. Space-based magnetic field measurements provided by the NOAA GOES and by the USAF DMSP are adequate but, again, limited in coverage. Ground-based magnetic field measurements available from the USGS network provide global warnings of geomagnetic storm activity, although localized regional warnings of geomagnetic storm intensity and duration would be enhanced through the use of international data from the INTERMAGNET consortium.

Aurora: Aurora formation begins with energetic solar particles following open magnetic field lines through the polar cusp into the Earth's polar regions. As the particles precipitate, they interact with atmospheric gas molecules and release large amounts of energy, some of which is in the visual spectrum. These visible emanations produce what is known as the Aurora Borealis and the Aurora Australis. Besides the visual aurora, the release of energy can cause scintillation within the polar ionosphere and ground-induced currents from the energized currents within the polar magnetic field. These conditions can change within seconds to minutes as the Earth experiences the sudden commencement of geomagnetic storms. Particle measurements available from the POES, MetOp and DMSP spacecraft are able to monitor the along-track location of the auroral boundary, as well as the auroral energy deposition from precipitating charged particles; the use of the DMSP UV scanning and limb sensors (SSUSI and SSULI) provides some off-track information as well. From these systems, coverage of the aurora domain is sufficient and provides continuous monitoring of auroral emissions and high-latitude scintillations.

Ionosphere: The ionosphere is a highly structured space weather domain, both vertically and horizontally. Ionospheric sounding data, available from the USAF DISS/NEXION network and other available international ionosondes, offer good vertical resolution, although the global coverage for these ground sensors is lacking. Powerful incoherent scatter radars can provide an excellent measurement of important ionosphere parameters and structure, but they too only cover a limited region and few exist worldwide. Although TEC measurements derived from ground-based GNSS receivers, such as the NASA JPL TEC and the NOAA CORS networks, can be extensive, this technique has poor vertical resolution and is currently limited to only land-based sites. The SSUSI and SSULI ultraviolet sensors on DMSP spacecraft provide some information, although the coverage is poor and the data latency from DMSP limits its stand-alone utility.

Likewise, the *in situ* sensors on DMSP provide information on ionospheric structure, but not continually, and only in a few local-time sectors. The planned COSMIC-2 system will provide unprecedented global coverage and sampling, although in this case, the horizontal resolution is limited. A preferred solution is to assimilate these diverse observational datasets into an environmental model which can then provide a global ionospheric specification. An example is the USAF GAIM model which is currently operational at AFWA and will soon be upgraded to a full physics-based version in 2014. The other aspect of this space weather domain is ionospheric scintillation which can have profound deleterious impacts on high-frequency radiowave communications and navigation, including precise geo-positioning. While the GPS radio occultation sensors on COSMIC-2 will be able to remotely sense GPS L-band scintillation, it is the availability of supporting observations, such as from the USAF C/NOFS and secondary sensors on board COSMIC-2, which will be able to monitor scintillation at other frequencies and aid in forecasting scintillation prior to their occurrence.

Upper Atmosphere: There are few operational assets available to sample the upper atmosphere at mesospheric (50 - 90 km) and thermospheric (90 - 1000 km) altitudes. The microwave radiometer on DMSP provides observations of mesospheric temperatures with limited altitude resolution and limited local time coverage. No observations of mesospheric winds are available operationally. Thermospheric neutral winds are observed by the Neutral Wind Meter (NWM), a single in situ sensor on the C/NOFS satellite that provides very limited altitude coverage and limited latitude coverage. However, visible light Doppler interferometers are under development with the capability to observe winds at a variety of thermospheric and mesospheric altitudes. Thermospheric neutral density profiles, neutral composition, and temperature observations are currently being provided for a range of altitudes (120 - 700 km), but with limited coverage in local time by the SSUSI and SSULI ultraviolet sensors on DMSP. The SENSE instrument, planned for operational demonstration in FY13, will carry an in situ sensor which provides neutral density, composition, and temperature at a fixed altitude (likely approximately 700 km). The proliferation of small in situ neutral density sensors on several orbit planes is one option for extending the local time coverage provided by the ultraviolet remote sensors on DMSP. These in situ sensors, however, are limited to altitudes above approximately 300 km, where satellite orbit lifetimes are prohibitively short due to effects of atmospheric drag. As in the case for the ionospheric domain, perhaps the best approach is to rely on atmospheric models that incorporate all available data, including calculated contributions from the coupled ionosphere.

Summary: The group's assessment of the ability of current and planned systems to satisfy documented space weather observing requirements is displayed in detail in Appendix 6. First, a detailed requirements analysis is presented for each of the six space weather domains, which includes an assessment of each observing system to measure the required environmental parameters within each of the domains (see Table 6-1 in Appendix 6). Second, detailed environmental parameter ratings for each of the five space weather phenomena are presented in terms of their impact/contribution on nowcasting, short-term forecasting, and long-term forecasting. These are evaluated for each relevant space environmental parameter, and then each parameter is prioritized as one of three factors: primary, secondary, or ancillary (see Table 6-2 in Appendix 6). A compilation of the detailed information from Appendix 6 is presented in Section 4.3 below.

4.3. Consolidated Analysis Results

A consolidated analysis of each space environmental parameter under each domain is presented in Table 3, below, which shows both the ability of current/planned instruments to meet observing requirements, as well as which environmental parameters are applicable to the five selected space weather phenomena at the three time scales (nowcasting, short-range forecasting, and long-range forecasting). The symbol and color assessments are directly linked to their respective environmental parameter ratings (EPR) from each domain worksheet for FY11 to FY22 (see Table 6-1 in Appendix 6). In terms of meeting requirements, those rated as "G" were the requirements that were mostly satisfied; "Y" were those requirements that were partly satisfied; "O" were those requirements that were addressed but with severe limitations; and "R" were those requirements that were not addressed or had severe limitations. As such, all were assigned the respective colors of green, yellow, orange, or red. Depending on the nature of the forecast requirements for a particular space weather scale, in some cases a "green" primary contributor (from Table 6-2 in Appendix 6) was sufficient to drive the overall roll-up assessment to green, whereas in other cases it was the ensemble of primary contributors that resulted in the overall roll-up color. Supporting contributors provided additional information for the roll-up, but these supporting contributors alone were not sufficient to drive the most favorable color. Ancillary contributors provided for the most part general situational awareness which represented at best a tertiary contribution to the overall score.

The top-level final roll-up chart presented in Table 4 provides a snapshot of the assessment to meet requirements to measure five key space weather phenomena. The symbol and color assessments are directly linked to their respective ratings for each environmental parameter used to monitor each phenomena from FY11 to FY22 (see Table 6-2 in Appendix 6), with a depiction of FY12, FY17, and FY21 as representative of years 0-3, 4-7, and 8-12, respectively. The ratings were directly traceable from this high level presentation to specific contributions provided by current and planned observational systems. Common to both Tables 3 and 4, part (A) illustrates the degradation of operational capability should these key systems be lost due to launch/system failure, budget cuts, or other reasons (i.e., the "worst case" scenario where none of the identified key replacement/upgrade observing systems are available). Likewise, part (B) depicts the sustainment of current capabilities over time if all these key systems are maintained or replaced (i.e., the "best case" scenario).

Table 3. Requirements Satisfaction by Space Weather Domain
(A) Worst Case

Table 3 is reproduced on the CD-ROM at the end of the Appendices.

Domain and Category	Requirements Satisfaction												Scoring Storms			Radio Blackouts			Radiation Storms			Enough Storms			Atmosph. Drag		
	JA11	JA12	JA13	JA14	JA15	JA16	JA17	JA18	JA19	JA20	JA21	JA22	Long	Short	Now	Long	Short	Now	Long	Short	Now	Long	Short	Now	Long	Short	Now
Solar Requirements																											
Solar EUV & UV Flux													B														B
Solar EUV & UV Imagery													B														
Solar Magnetic Field													B												B		B
Solar Radio Emissions (Total & spectral flux)													B			B			B			B					B
Solar Radio Burst (Location, Time, Polarization)													B			B			B			B					
Solar Imagery IR and Optical													B														
Solar Coronagraph													B			B			B			B			B		B
Solar X-Ray Flux (total and discrete frequency)													B			B			B			B			B		
Solar X-Ray Imagery													B														
Off-Angle Solar Imagery (possibly L5)													B														
Helioseismology																											
Heliosphere / Solar Wind																											
Solar Wind: 3D Magnetic Field Components @ L1													B						B		B	B	B	B	B	B	B
SW Plasma Components: Comp, Den & Temp @ L1													B						B			B	B	B	B	B	B
Solar Wind: Speed and Direction @ L1													B									B					B
Sun-Earth Line based Heliospheric Imagery													B									B					
Off-Angle Heliospheric Imagery (L4 or L5)													B									B					
Solar Wind Radio Emission													B									B					
Solar High Energy Protons and Cosmic Rays													B									B			B		
Solar Relativistic Electrons @ L1 or L2													B									B					
Off-Angle Solar Wind/Mag - in-situ (possibly L5)																											
Magnetosphere																											
Energetic Charged Particles: Energy & Flux																						B	B	B			B
Medium Charged Particles: Energy & Flux																B						B	B	B			B
Trapped Radiation: Protons, Electrons, Waves																B						B	B	B			
Supra-thermal Charged Particles																						B					
Magnetic Field - in-situ (GEO & LEO)																											
Geomagnetic Field - Surface																						B			B		

Space Weather Observing Systems Capability Assessment

Aurora

Parameter	FY11	FY12	FY13	FY14	FY15	FY16	FY17	FY18	FY19	FY20	FY21	FY22
Auroral Boundary	X	X	X	X	X	X	X	X	X	X	X	X
Auroral Energy Deposition	X	X	X	X	X	X	X	X	X	X	X	X
Auroral Emissions & Imagery	L	L	L	L	L	U	U	U	U	U	U	U
Precipitating Charged Particles	X	X	X	X	X	X	X	X	X	X	X	X

Ionosphere

Parameter	FY11	FY12	FY13	FY14	FY15	FY16	FY17	FY18	FY19	FY20	FY21	FY22
Ionospheric Scintillation	L	L	L	L	L	L	L	L	L	L	U	U
Plasma Density Fluctuations	L	L	L	L	L	L	L	L	L	L	L	(L)
Plasma Temperatures (Te & Ti)	L	L	L	L	L	L	L	L	L	L	L	(L)
Ionospheric Char: Layer Height & Density	L	L	L	L	L	L	L	L	L	L	L	U
Energetic Ions (D-region absorption)	L	L	L	L	L	L	L	L	L	L	L	U
Total Electron Content	L	L	L	L	L	L	L	L	L	L	L	L
Electric Field	X	X	X	X	X	X	X	X	X	X	X	(L)
D-Region Absorption						X						X
Electron Density Profile			U			U	U	U	U	U	U	(U)

Upper Atmosphere

Parameter	FY11	FY12	FY13	FY14	FY15	FY16	FY17	FY18	FY19	FY20	FY21	FY22
Mesospheric Temperature	U	U	U	U	U	U	U	U	U	U	U	(U)
Mesospheric Winds (Speed & Direction)												
Neutral Winds (Speed & Direction)	U	U	U	U	U	U	U	U	U	U	U	(U)
Neutral Density, Composition & Temperature	L	L	L	U	U	U	U	U	U	U	U	(U)
Neutral Density Profile	L	L	L	L	L	U	U	U	U	U	U	(U)

Last Update: 3 April, 2012 (Worst Case)

Table 3. Requirements Satisfaction by Space Weather Domain (continued)
(B) Best Case

Table 3 is reproduced on the CD-ROM at the end of the Appendices.

Domain and Category	Requirements Satisfaction	Geomag Storms			Radio Blackouts			Radiation Storms			Ionosph Storms			Atmosph Drag		
		LOW	MED	HIGH	LOW	MED	HIGH	LOW	MED	HIGH	LOW	MED	HIGH	LOW	MED	HIGH
Solar Requirements																
Solar EUV & UV Flux																
Solar EUV & UV Imagery																
Solar Magnetic Field																
Solar Radio Emission (Total & spectral flux)																
Solar Radio Burst (Location, Type, Polarization)																
Solar Imagery IR and Optical																
Solar Coronagraph																
Solar X-Ray Flux (total and source frequency)																
Solar X-Ray Imagery																
Off-Angle Solar Imagery (possibly L5)																
Helioseismology																
Heliosphere / Solar Wind																
Solar Wind 3D Magnetic Field Components @ L1																
SW Plasma Components: Comp, Den & Temp @ L1																
Solar Wind Speed and Direction @ L1																
Sun-Earth Line based Heliospheric Imagery																
Off-Angle Heliospheric Imagery (L4 or L5)																
Solar Wind Radio Emissions																
Solar High Energy Protons and Cosmic Rays																
Solar Relativistic Electrons @ L1 or L5																
Off-Angle Solar Wind/Mag - In-situ (possibly L5)																
Magnetosphere																
Energetic Charged Particles: Energy & Flux																
Medium Charged Particles: Energy & Flux																
Trapped Radiation - Proton, Electron, Waves																
Supra-thermal Charged Particles																
Magnetic Field - In-situ (GEO & LEO)																
Geomagnetic Field - Surface																

Space Weather Capability Matrix

Aurora
- Auroral Boundary
- Auroral Energy Deposition
- Auroral Emissions & Imagery
- Precipitating Charged Particles

Ionosphere
- Ionospheric Scintillation
- Plasma Density Fluctuations
- Plasma Temperatures (Te & Ti)
- Ionospheric Char: Layer Height & Density
- Energetic Ions (D-region absorption)
- Total Electron Content
- Electric Field
- D-Region Absorption
- Electron Density Profile

Upper Atmosphere
- Mesospheric Temperature
- Mesospheric Winds (Speed & Direction)
- Neutral Winds (Speed & Direction)
- Neutral Density, Composition & Temperature
- Neutral Density Profile

(Columns: FY11, FY12, FY13, FY14, FY15, FY16, FY17, FY18, FY19, FY20, FY21, FY22, and Long / Short / Now time ranges)

Last Update: 3 April 2012 (Best Case)

LEGEND		
	X	Satisfactory (Fully or nearly meets requirements)
Meets requirements	L	Usable with limitations (Limited in capability and/or coverage)
Meets most Requirements	U	Usable with severe limitations (Limited in capability and/or coverage)
Meets some requirements	( )	Asset may not be available due to operational status, program funding, etc.
	[O]	No capability
Fails to meet requirements	#	Environmental parameter is applicable to the space weather phenomenon

When consolidating these requirements and considering the ability of the current/planned systems to monitor the five key space weather phenomena previously discussed, high-level impacts tied to few key systems become apparent. It is particularly noteworthy that the addition of planned replacements or new systems maintains or incrementally upgrades our current capabilities; as such, none of these planned/replacement systems meet all requirements. Perhaps even more importantly, this demonstrates the significant degradation in current capability should these planned/replacement systems not reach operational status. In other words, the Nation is at risk of losing critical capabilities that have significant economic and security impacts should these key space weather observing systems fail to be maintained and replaced. Considering the rapidly growing dependency on space-based and space-enabled systems, which have permeated most facets of modern society, space weather observing and forecasting capabilities used to mitigate potential impacts will become even more critical in the future.

Table 4. Requirements Satisfaction by Phenomena

(A) Worst Case

Timeline (years)	Nowcasts (Current Conditions)			Short-term Forecasts (minutes to hours)			Long-term Forecasts (hours to days)		
	0-3	4-7	8-12	0-3	4-7	8-12	0-3	4-7	8-12
Geomagnetic Storms	G	G	G	Y (3) O	O	O	Y (4) O	O	O
Radio Blackouts	G	G	G	O	O	O	Y	Y	Y
Solar Radiation Storms	G	G	G	Y (2) R	R	R	O	O	O
Ionospheric Storms	Y (1) O	O	O	Y (3) O	O	O	Y (4) R	R	R
Atmospheric Drag	Y (1) O	O	O	Y (3) O	O	O	Y (4) R	R	R

Substantial degradation over time if systems aren't sustained or replaced

G	Meets Requirements
Y	Limited Capability
O	Severely Limited Capability
R	Fails to Meet Requirements

(1) Reduced DMSP coverage from two to one orbits
(2) Loss of relativistic electron data SOHO.
(3) Uncertainty of solar wind data from L1 to replace ACE.
(4) Uncertainty of getting a space-based coronagraph to replace SOHO and STEREO data.

Table 4. (Continued)
(B) Best Case

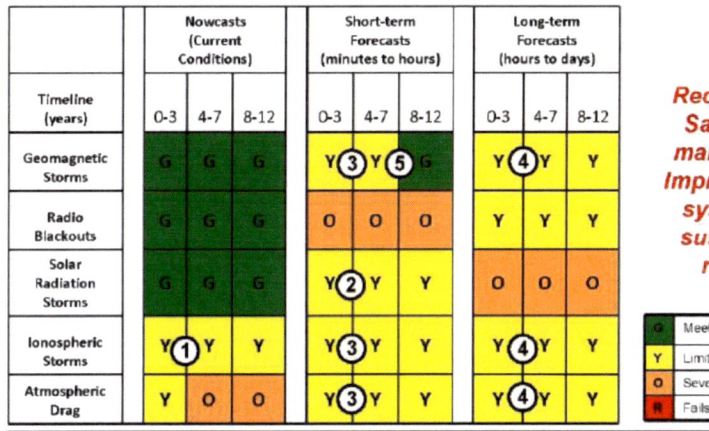

	Nowcasts (Current Conditions)			Short-term Forecasts (minutes to hours)			Long-term Forecasts (hours to days)		
Timeline (years)	0-3	4-7	8-12	0-3	4-7	8-12	0-3	4-7	8-12
Geomagnetic Storms	G	G	G	Y ③ Y	⑤	G	Y ④ Y		Y
Radio Blackouts	G	G	G	O	O	O	Y	Y	Y
Solar Radiation Storms	G	G	G	Y ② Y		Y	O	O	O
Ionospheric Storms	Y ① Y		Y	Y ③ Y		Y	Y ④ Y		Y
Atmospheric Drag	Y	O	O	Y ③ Y		Y	Y ④ Y		Y

Requirements Satisfaction maintained or Improved if key systems are sustained or replaced

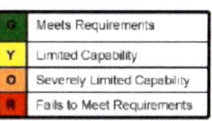

G	Meets Requirements
Y	Limited Capability
O	Severely Limited Capability
R	Fails to Meet Requirements

(1) COSMIC-2 deployed
(2) Relativistic electron data from SOHO are obtained.
(3) Solar wind data from L1 to replace ACE is obtained.
(4) Space-base coronagraphs on SOHO and STEREO are replaced.
(5) Advanced plasma sensor on DSCOVR follow-on obtained.

5. KEY FINDINGS

In performing the assessment of current and planned space weather observing systems and evaluating the ability of those systems to meet documented requirements, the JAG/SEGA made several key findings summarized below.

5.1. Summary of Key Findings

In performing its assessment, the JAG/SEGA reached the following key findings:

- A judicious mix of space-based and ground-based observing systems are currently used and needed to support operational space weather services.
  - o The huge volume of the space environment means that

even with the dozens of observing systems now used, there are still limited observational data to produce space weather forecasts.

- Research observing systems provide important data used to advance science; many of those also provide timely data and are used to support operational space weather services.
  - o Several NASA heliospheric research missions will reach end-of-life within the next 10 years.
- Several NOAA and DoD space-based operational systems are scheduled to be replaced over the next 10 years subject to available funding.
- While ground-based systems are in important component to the space weather mission, sparse coverage limits their utility in meeting operational requirements.
- A number of foreign space-based and ground-based capabilities are used to help meet U.S. operational space weather needs.
  - o More are available and provide the potential for future use.
  - o While foreign data sources can provide additional capability, the economic and national security interests of the United States dictate that the nation not rely exclusively on foreign assets to conduct the critical space weather mission.
- Most unexploited data sources (foreign and domestic) are not currently used due to lack of reliable or timely access, excessive expense, policy/security restrictions, or other practical reasons. Also, these data sources offer secondary capabilities that cannot replace key, primary systems. Nevertheless, many offer added value that could incrementally improve forecasting and should be used when feasible and cost-effective.
- While space-based and ground-based observing systems are a critical components needed to meet operational requirements, they are inextricably linked to other parts of the space weather architecture (such as models and other space weather forecasting capabilities), and thus should not be considered alone when assessing our ability to meet requirements.

6. SUMMARY

As part of the 2010 NASA Authorization Act, Congress asked OSTP to submit a report to the appropriate committees of Congress that (1) details the current data sources, both space- and ground-based, that are necessary for space weather forecasting; and (2) details the space- and ground-based systems that will be required to gather data necessary for space weather forecasting for the next 10 years. In turn, OSTP requested the assistance of the Office of the Federal Coordinator for Meteorological Services and Supporting Research (OFCM)-sponsored National Space Weather Program Council (NSWPC). The NSWPC immediately stood up the Joint Action Group for Space Environmental Gap Analysis (JAG/SEGA) to perform the assessment to provide the requested information to OSTP. The JAG/SEGA, comprised of 25 individuals from 15 different Federal organizations, analyzed current and planned space weather observing systems and assessed their ability to meet existing requirements formally documented by DOC (NOAA), DoD, and NASA. Interim results were presented to the NSWPC on August 2, 2011, with OSTP and OMB present as observers. This report constitutes the final results, which includes results from the JAG's assessment.

As the Sun approaches its next peak of solar activity, expected in 2013, our Nation faces multiplying uncertainties from increasing reliance on technologies for communications, navigation, security, and other activities, many of which both underpin our national infrastructure and economy and are vulnerable to the effects of space weather. Our Nation also faces increasing exposure to space-weather-driven human health risks as trans-polar flights and space activities, including space tourism and space commercialization, increase. Therefore, for the benefit of our national security, economy, and public welfare, it is more important than ever to ensure that the Nation's space weather observing and forecasting capabilities are supported and maintained.

ABBREVIATIONS AND ACRONYMS

3D	3 Dimensional
A3O-W	Air Force Directorate of Weather
ACE	Advanced Composition Explorer
ACE/MAG	ACE Magnetometer
AFRL	Air Force Research Laboratory

AFSPC	Air Force Space Command
AFWA	Air Force Weather Agency
AIA	Atmospheric Imaging Assembly
AMPERE	Active Magnetosphere and Planetary Electrodynamics Response Experiment
AMSU	Advanced Microwave Sounding Unit
ATMS	Advanced Technology Microwave Sounder
AU	Astronomical Unit
BDD	Burst Detector Dosimeter
cm	centimeter(s)
CME	Coronal Mass Ejection
C/NOFS	Communications/Navigation Outage Forecast System
CORS	Continuously Operating Reference Stations
COSMIC	Constellation Observing System for Meteorology, Ionosphere, and Climate
CSA	Canadian Space Agency
CTIP	Cubesat Tiny Ionospheric Photometer
CSW	Committee for Space Weather
CXD	Combined X-ray Dosimeter
DISS	Digital Ionospheric Sounding System
DMSP	Defense Meteorological Satellite Program
DNB	Day-Night Band
DOC	Department of Commerce
DoD	Department of Defense
DOE	Department of Energy
DOS	Department of State
DOT	Department of Transportation
DSCOVR	Deep Space Climate Observatory
DSN	Deep Space Network
EDP	Electron Density Profile
EHIS	Energetic Heavy Ion Sensor
EIT	Extreme ultraviolet Imaging Telescope
EOL	End of Life
EPR	Environmental Parameter Ratings
EPS-HES	Energetic Particle Sensor - High Energy Sensor
ESA	European Space Agency
ESP	Energetic Spectrometer for Particles
EUMETSAT	European Organisation for the Exploitation of

	Meteorological Satellites
EUV	Extreme Ultraviolet
EUVI	Extreme UltraViolet Imager (LMSAL)
eV	electron Volt
EVE	Extreme Ultraviolet Variability Experiment
EXIS	EUV and X-ray Irradiance Sensors
FAA	Federal Aviation Administration
FOC	Full Operational Capability
FY	Fiscal Year
GAIM	Global Assimilation of Ionospheric Measurements
GCR	Galactic Cosmic Rays
GEO	Geosynchronous Earth Orbit
GNSS	Global Navigational Satellite System
GOCE	Gravity field and steady-state Ocean Circulation Explorer
GOES	Geostationary Operational Environmental Satellites
GOES NOP	GOES N-O-P Series Satellites
GOES MAG	GOES Magnetometer
GOES-R	GOES - R series satellites
GOES-R/MAG	GOES-R Magnetometer
GONG	Global Oscillation Network Group
GONG/FT	GONG Fourier Tachometer
GPS	Global Positioning System
GPSRO	GPS Radio Occultation
GRACE	Gravity Recovery and Climate Experiment
GSFC	Goddard Space Flight Center
GTO	Geosynchronous Transfer Orbit
HASDM	High Accuracy Satellite Drag Model
HEO	Highly Elliptical Orbit
HEPAD	High Energy Particle Detector
HF	High Frequency
HMI	Helioseismic and Magnetic Imager
HOPE	Helium Oxygen Proton Electron
IOC	Initial Operational Capability
IMF	Interplanetary Magnetic Field
INTERMAGNET	International Real-time Magnetic Observatory Network
ISOON	Improved Solar Observing Optical Network
IT	Information Technology

IVM	Ion Velocity Monitor
JAG	Joint Action Group
JAG/SEGA	Joint Action Group for Space Environmental Gap Analysis
JPL	Jet Propulsion Laboratory
JPSS	Joint Polar Satellite System
keV	kilo electron Volt
kHz	kiloHertz
km	kilometer(s)
L1	Earth-Sun Lagrangian point 1
L2	Earth-Sun Lagrangian point 2
L4	Earth-Sun Lagrangian point 4
L5	Earth-Sun Lagrangian point 5
LANL	Los Alamos National Laboratory
LASCO	Large Angle and Spectrometric Coronagraph
LBHl/LBHs ratio	Lyman-Birge-Hopfeld auroral i/s ratio
LEO	Low Earth Orbit
LEPS	Low Energy Particle Sensor
LOS	Line of Sight
LWS	Living With a Star
MagEIS	Magnetic/electric Field Instrument Suite
MDI	Michelson Doppler Imager
MEPED	Medium Energy Proton and Electron Detector
MetOp	Meteorological Observation satellite (EUMETSAT)
MeV	Mega electron Volt
MF/HF	Medium Frequency/High Frequency
MHz	Megahertz
MLS	Microwave Limb Sounder
MPA	Magnetospheric Plasma Analyzer
MPS-HI	Magnetospheric Particle Sensor - High
MPS-LO	Magnetospheric Particle Sensor -Low
NASA	National Aeronautics and Space Administration
NDP	Neutral Density Profile
NESDIS	National Environmental Satellite Data and Information Service
NEXION	Next Generation Ionosonde
NGA	National Geospatial-Intelligence Agency
NGS	National Geodetic Survey
NOAA	National Oceanic and Atmospheric Administration

nP	nano Pascals
NPOESS	National Polar-orbiting Operational Environmental Satellite System
NPP	NPOESS Preparatory Project
NRC	National Research Council
NRCC	National Research Council of Canada
NRL	Naval Research Laboratory
NRT	Near Real Time
NSF	National Science Foundation
NSO	National Solar Observatory
NSWPC	National Space Weather Program Council
nT	nano Tesla
NWM	Neutral Wind Meter
NWS	National Weather Service
OFCM	Office of the Federal Coordinator for Meteorologica Services and Supporting Research
OLS	Operational Linescan System
OMB	Office of Management and Budget
OSIRIS	Optical Spectrograph and InfraRed Imager System
OSTP	Office of Science and Technology Policy
PLP	Planar Langmuir Probe
POES	Polar Operational Environmental Satellite
RBSP	Radiation Belt Storm Probe
RE	Earth Radii
REPT	Relativistic Electron Proton Telescope
RIMS	RSTN Radio Interference Measurement Set
RS	Solar Radii
RSTN	Radio Solar Telescope Network
SABER	Sounding of the Atmosphere using Broadband Emission Radiometry
SABRS	Space Atmospheric Burst Reporting System
SATCOM	Satellite Communications
SBUV	Solar Backscatter Ultraviolet
S/C	Spacecraft
SCI	Sensitive Compartmented Information
SCINDA	Scintillation Network Decision Aid
SDO	Solar Dynamics Observatory
SECCHI	Sun Earth Connection Coronal and Heliospheric Investigation

SEM	Space Environmental Monitor
SEM-2	Space Environmental Monitor - 2
SEM-N	Space Environmental Monitor - Next
SENSE	Space Environmental Nanosat Experiment
SEON	Solar Electro-Optical Network
SEP	Solar Energetic Particle
SGPS	Solar and Galactic Proton Sensor
SIESS	Space Environment In-Situ Suite
SIS	ACE Solar Isotope Spectrometer
SMC	Space and Missile Systems Center
SMEI	Solar Mass Ejection Imager
SOHO	Solar and Heliospheric Observatory
SOON	Solar Observing Optical Network
SOPA	Synchronous Orbit Particle Analysis
sr	steradians
SSAEM	Space Situational Awareness Environmental Monitoring
SSIES	Special Sensors-Ions, Electrons, and Scintillation
SSJ	Special Sensor J
SSM	Special Sensor Magnetometer
SSMIS	Special Sensor Microwave Imager Sounder
SSULI	Special Sensor UV Limb Imager
SSUSI	Special Sensor UV Spectrographic Imager
STC	Science and Technology Corporation
STEREO	Solar TErrestrial RElations Observatory
SuperDARN	Super Dual Auroral Radar Network
SUVI	Solar Ultraviolet Imager
SWACI	Space Weather Applications Center - Ionosphere
SWEPAM	Solar Wind Electron Proton Alpha Monitor
SWPC	Space Weather Prediction Center
SXI	Solar X-Ray Imager
TEC	Total Electron Content
THEMIS	Time History of Events and Macroscale Interactions
TIDI	TIMED Doppler Imager
TIMED	Thermosphere Ionosphere Mesosphere Energetics and Dynamics
UHF	Ultra High Frequency
U.S.	United States
USAF	United States Air Force

USGS	United States Geological Survey
USNDS	U.S. Nuclear Detonation (NUDET) Detection System
UV	Ultraviolet
UVI	UV Imager
VHF	Very High Frequency
VIIRS	Visible Infrared Imaging Radiometer Suite
WINCS	Wind Ion Neutral Composition Suite
XRS	Solar X-Ray Sensor

ACCOMPANYING CD-ROM T.O.C.

REPORT ON SPACE WEATHER OBSERVING SYSTEMS: CURRENT CAPABILITIES AND REQUIREMENTS FOR THE NEXT DECADE

APPENDICES

(OFFICE OF THE FEDERAL COORDINATOR FOR METEOROLOGICAL SERVICES AND SUPPORTING RESEARCH)

INDEX

D

E

S

T